DISCOVERING POSTMODERN COSMOLOGY

Discoveries in Dark Matter, Cosmic Web, Big Bang, Inflation, Cosmic Rays, Dark Energy, Accelerating Cosmos

JEROME DREXLER

Universal Publishers
USA • 2008

DISCOVERING POSTMODERN COSMOLOGY:
Discoveries in Dark Matter, Cosmic Web, Big Bang,
Inflation, Cosmic Rays, Dark Energy, Accelerating Cosmos

Universal Publishers
Boca Raton, Florida • USA
2008

ISBN-10: 1-599-987-X
ISBN-13: 978-1-59942-987-8

www.universal-publishers.com

*This book is dedicated to Sylvia, my wife,
best friend, and lifelong partner.*

*"Leave the beaten track occasionally and dive into the woods.
Every time you do so you will find something you have never seen
before. Follow it up, explore all around it, and before you know
it, you will have something to think about to occupy your mind.
All really big discoveries are the result of thought."*

~ Alexander Graham Bell ~

*"We are to admit no more causes of natural things than such are
both true and sufficient to explain their appearances."*

~ Issac Newton's version of Occam's razor ~

"Make everything as simple as possible, but not simpler."

~ Albert Einstein ~

*"It is dangerous to be right in matters on which the established
authorities are wrong."*

~ Voltaire ~

PREFACE

Learn how a world-class inventor-scientist is currently tackling the greatest scientific mysteries of the universe — and succeeding. This book could revolutionize how research in cosmology will be conducted. Follow the progress as new astronomical findings are accumulated from NASA and others to help substantiate and verify this *postmodern cosmology* model.

"Seven Quick Cosmic Discoveries Create New Cosmology and Can Be Examined Here" could have been the 19th century title for this book. This is the first book to address all seven thought-to-be-unsolved cosmic phenomena mysteries shown in this book's subtitle. It is also the first to offer plausible explanations for all seven of the *unsolved* mysteries. When accepted by the cosmologists, this research could provide a viable baseline to jump-start debate on a standard model for *postmodern cosmology*.

This book weaves together seven cosmic phenomena thereby forming a fabric for *postmodern cosmology*. It was written for open-minded cosmologists, astronomers, astrophysicists, physicists, engineers, students, enthusiasts and those at NASA, NSF, DOE and ESO who want to understand *postmodern cosmology*. This new cosmology is based upon the nature of these seven cosmic phenomena and upon their relationships and linkages.

Is the dark matter of the universe comprised of ultra-high-energy relativistic protons? If it is, seven major cosmic mysteries have been solved. Is this book science fiction or does it measure up to the scientific contributions of the book of Nicolaus Copernicus, *On The Revolutions Of Heavenly Spheres*. Which is it?

This book starts out with the premise that the dark matter of the universe is comprised of ultra-high-energy relativistic protons. It then uses this simple premise, existing astronomical data, and the laws of physics to explain five major cosmic mysteries in a plausible manner. By accomplishing this five times, the premise itself becomes plausible. This procedure is explained as follows:

The only plausible explanation for the *accelerating expansion of the universe*, announced to date, requires that the dark matter of the universe be comprised of ultra-high-energy relativistic protons (see Chapter 18).

The only plausible explanation for the *Big Bang satisfying the Second Law of Thermodynamics*, announced to date, requires that dark matter be comprised of ultra-high-energy relativistic protons (see Chapter 12).

The only announced plausible explanation for *ultra-high-energy cosmic ray protons* with energies above 60 EeV bombarding Earth's atmosphere requires that dark matter be comprised of ultra-high-energy relativistic protons (see Chapters 29 and 31).

The only announced plausible explanation for *cosmic inflation* requires that dark matter be comprised of ultra-high-energy relativistic protons (see Chapters 31 and 34).

The only announced plausible explanation for the *cosmic web* requires that dark matter be comprised of ultra-high-energy relativistic protons (see Chapter 33).

If these five explanations are indeed plausible, it is highly probable that the dark matter of the universe is comprised of ultra-high-energy relativistic protons.

My first book, *How Dark Matter Created Dark Energy and the Sun,* and I appeared on the astro-cosmology scene together on Dec.15, 2003. Although I had been CEO and chief scientist of a Silicon Valley NASDAQ company for decades, no one in cosmology had ever heard of me.

I had built a home library of 20 books in cosmology, astronomy, and astrophysics beginning 1999 and spent my spare time studying them, not just reading them. After awhile I began to realize that my authors did not have convincing arguments regarding the nature of *dark matter*. Since Cold

Dark Matter had been proposed in 1984, that candidate did not look very promising with no progress after 16 years.

Then it occurred to me that if no one seems to know what *dark matter* is, maybe that's the new scientific challenge for which I have been looking. Although I have been granted 76 US patents and have created two high-tech companies with my inventions, discovering the nature of dark matter seemed like the challenge of a lifetime.

By 2002, I decided I was sure that *dark matter* was comprised of galaxy-orbiting relativistic protons and I proceeded to enter the field via several approaches that looked promising. Einstein's 1905 Special Theory of Relativity was invaluable. Johannes Kepler encouraged me in that, as a mathematician, he was able to interpret Tycho Brahe's astronomical data and develop Kepler's laws, which were beyond the astronomers of the day. Occam's razor logic, which I had admired and used all my life, was there to help me. Finally, there was my early interest in cryptography and my Bell Labs course in Information Theory that might give me an edge in data analysis.

I assembled the material for my astro-cosmology book beginning in the spring of 2003. Then to test the theory, I gave a 32-slide Powerpoint presentation to two astronomy/astrophysics professors at the University of California, Santa Cruz. It gave me added confidence. I gave up my job as CEO and chief scientist of Drexler Technology Corporation, a NASDAQ company in Silicon Valley, on Sept. 1, 2003. My 156-page paperback book, *How Dark Matter Created Dark Energy and the Sun,* was published Dec.15, 2003.

On April 22, 2005, I published online a 19-page scientific paper on the Cornell University Library's arXiv.gov website as e-Print No. astro-ph/0504512 entitled "Identifying Dark Matter through the Constraints Imposed by Fourteen Astronomically Based 'Cosmic Constituents'".

Upon discovering that my dark matter theory was able to solve over 15 unsolved cosmic mysteries, I wrote a 295-page paperback sequel. The book, entitled *Comprehending and Decoding the Cosmos*[8], was published May 22, 2006.

This 2006 book also discloses the roles and functions of Relativistic-Proton Dark Matter in creating the cosmic web, spiral galaxies, stars, starburst galaxies and ultra-high-energy cosmic rays.

About a month later several astro-cosmology news items were published that provided support for my theories. I issued my first *cosmic newswire* about them on the Internet on June 27, 2006. Positive news from Russia led to a second *cosmic newswire* on July 17. A report of extreme ultraviolet emission from galaxy clusters led to a third one on Aug.21. The Sept. 5 *cosmic newswire* was entitled "Dark Matter's

Identity Revealed by Deciphering 14 Cosmic Clues". During 2006, eight of these *cosmic newswires* were issued.

The opportunities to issue encouraging *cosmic newswires* supporting my theories continued unabated for 20 months following the publishing of the May 2006 book, resulting in the issuance of thirty-one relevant *cosmic newswires* through January 2008.

My scientific paper, "A Relativistic-Proton Dark Matter Would Be Evidence the Big Bang Probably Satisfied the Second Law of Thermodynamics"[13], was published online on February 15, 2007. It solidified my research base, played a key role in developing *postmodern cosmology*, and encouraged me to write this 2008 book.

The 2007 scientific paper and the collection of thirty-one *cosmic newswires* (some updated and/or edited) provide overwhelming evidence that the key astronomical and analytical discoveries have been made to support my theories of the following seven cosmic phenomena: *dark matter*, *cosmic web*, the *Big Bang*, *cosmic inflation*, energetic *cosmic rays*, *dark energy*, and the *accelerating universe*. They provide the scientific material and substance from which the fabric of *postmodern cosmology* is woven.

CONTENTS

Discovering Postmodern Cosmology

Discoveries in Dark Matter, Cosmic Web, Big Bang, Inflation, Cosmic Rays, Dark Energy, Accelerating Cosmos

INTRODUCTION

Since the thirty-one cosmic newswires you are about to read rely on and apply to the Relativistic-Proton Dark Matter theory as well as to the postmodern Big Bang cosmology theory, let us begin by introducing these theories.

The author believes that the galaxy-orbiting relativistic proton appears to have the necessary characteristics of the long-sought dark matter particles, which are estimated by most scientists to comprise about 83 percent of the total mass of the universe. Relativistic protons do have the required mass and the required difficulty of detection and can transform themselves into hydrogen, the principal matter of galaxies, by combining with electrons created through pion producing collisions and pion decay. (See Appendix A regarding the scientific community's long-held objections to any proton dark matter and the author's response.)

Therefore, relativistic protons are capable of forming (1) galaxies and their dark matter halos, (2) galaxy clusters and their distributed dark matter, (3) the cosmic web, the long, large, filamentary dark matter that crisscrosses the cosmos, and (4) newborn stars and igniting their hydrogen fusion reactions.

However, for this Relativistic-Proton Dark Matter theory to become widely accepted, there also should be astronomical evidence of multitudinous relativistic protons within the spheroidal dark matter halo surrounding the Milky Way and other spiral galaxies. The author believes that the energetic cosmic ray relativistic protons bombarding Earth every day go a long way toward providing such astronomical evidence.

Cosmologists, who are known to be scientifically conservative, have not yet accepted the author's explanation of the nature of dark matter, the cosmic web, dark energy, the Big Bang, the accelerating expansion of the universe, the energy source(s) for ultra-high-energy cosmic ray protons, or cosmic inflation. Hopefully, this book will provide sufficient scientific evidence to convince some of the cosmologists of the validity or plausibility of some of these explanations.

In order to strengthen his case for Relativistic-Proton Dark Matter with the cosmologists, the author devised in 2005 a second independent approach, beyond his 2003 published approach, to confirm the identity of dark matter. Since dark matter represents about 83 percent of the mass of the universe, it is omnipresent and should have an influence on or a relationship with a number of celestial bodies. Such relationships might be used to identify dark matter, he felt.

Meanwhile, the vast majority of research conducted on dark matter by physicists has had to do with trying to identify the particles that comprise dark matter or to determine their gravitational effect on star rotation curves in spiral galaxies. This primarily inward-looking approach to identify the particle composition of a medium is known as *reductionism,* which is a procedure or theory that reduces or attempts to reduce complex data or phenomena into simple elements.

Reductionism does not always work well in physics. Many times simple entities or particles can form complex forms or combinations that have characteristics seemingly unrelated to the characteristics of the original simple entities. A hurricane is one well-known example of complex behavior whose characteristics cannot be predicted by an analysis of all the known simple entities involved in its makeup. Thus, the reductionism approach does not explain or predict the nature of a hurricane.

An alternative to the inward-looking reductionism is an outward-looking, cosmological-like approach that the author has developed and designated *relationism,* where a phenomenon such as the dark matter can be analyzed and categorized in terms of its various relationships. *Dark matter relationism,* described in the author's 2006 book, is used to provide additional evidence that Relativistic-Proton Dark Matter is the dark matter of the universe.

This 2008 book provides a third independent approach to build the case for Relativistic-Proton Dark Matter. It argues that the Big Bang, which occurred at the beginning of time, must have satisfied the Second Law of Thermodynamics. Thus, immediately after the extremely high energy Big Bang, the entropy (disorder) of the universe would be at the lowest level it would reach throughout all time. This could be achieved by the Big Bang firing out, in all directions, high-velocity ultra-high-energy (UHE) relativistic protons and helium nuclei in the well-known nuclei ratio near 12:1.

Such a Big Bang, characterized by a *violent radial dispersion* of ultra-high-energy relativistic nuclei, would be highly efficient and could create very high usable energy and have very low entropy, and might be designated a *Relativistic Big Bang*. This Big Bang concept is fundamental to the author's Relativistic-Proton Dark Matter theory and to his postmodern Big Bang cosmology.

Postmodern cosmology and postmodern Big Bang cosmology mean the same thing. Their explanation begins with requiring that the Big Bang satisfy the Second Law of Thermodynamics, which essentially requires that the vast majority of mass/energy produced by the Big Bang be in the form of relativistic protons and helium nuclei. Dark matter represents 83 percent of the mass of the universe, and was produced by the Big Bang. Therefore dark matter would have to be comprised of relativistic protons and helium nuclei.

Over 99 percent of the mass of the universe is hydrogen and helium in the atomic ratio of about 12:1. The Big Bang created all the mass of the universe, with almost all consisting of relativistic protons and helium nuclei in the atomic ratio not far from 12:1. Since 83 percent of the mass of the universe is dark matter, it follows that dark matter should be comprised of relativistic protons and helium nuclei in a ratio not far from 12:1. The name Relativistic-Proton Dark Matter is defined by this specific mix of protons and helium nuclei, which are both baryons. In 2008, the more accurate name, Relativistic-Baryon Dark Matter began to be used.

In this book we will develop the only publicly announced plausible explanations for *dark matter*, the *cosmic web*, the *Big Bang*, the *accelerating expansion of the universe*, for the existence of the *ultra-high-energy cosmic-ray protons*, for *cosmic inflation*, and insights into the nature of *dark energy*.

Relativistic-Proton Dark Matter is also used to explain, in the author's 2006 book[8], how the first stars were created, why a spiral galaxy creates blue stars in its spiral arms, why starburst galaxies create blue stars in their cores and have star formation rates fifty times higher than that of a spiral galaxy, and why there is a different form of dark matter around small galaxies as compared to dark matter around groups of galaxies within normal galaxy clusters.

Postmodern cosmology requires a change of focus, for physicists and cosmologists, onto relationism rather than only on reductionism. It requires that the universe be looked upon as a complex, orderly, self-inclusive system in space and time comprised of interrelated cosmic entities, rather than simply a collection of unrelated cosmic entities.

A number of the chapters provide published research results from other scientists casting doubt on the existence of Cold Dark Matter[4], which remains intangible 23 years after it was proposed. In contrast, the author is not aware of published scientific research data casting doubt on the existence of the 4-year-old Relativistic-Baryon Dark Matter.

Therefore, it seems as if a dark-matter *fork in the road* has appeared on the horizon. In order for cosmologists to make significant progress on *postmodern cosmology* they will have to decide which dark-matter fork to take. Study of this book could be helpful in analyzing the alternatives.

Some of the early history of dark matter research can be found in the first four pages of Chapter 12.

Appendix A: Scientists' long-held objections to proton-based dark matter are presented along with this author's responses. It should be read by professional astronomers, astrophysicists, and cosmologists.

Appendix B: Presented here are 18 selected pages involving six important references from J. Drexler's December 2003 book *How Dark Matter Created Dark Energy and the Sun.* It is the first book of Drexler's astro-cosmology trilogy.

Appendix C: Presented here are Chapters 44-48, 50 and 56, the Epilogue, from J. Drexler's May 2006 book *Comprehending and Decoding the Cosmos.* It is the second book of Drexler's astro-cosmology trilogy. These chapters are designated Chapters 2006-44, 2006-45, etc.

A 20-page glossary and 47 references are provided.

CHAPTER 1

COSMIC NEWSWIRE #1

More Evidence That Dark Matter of Universe Is Not Cold Dark Matter

LOS ALTOS HILLS, Calif., June 27, 2006 (AScribe Newswire) -- The current mainstream theory of the dark matter of the universe, in the United States, is that dark matter is cold and collisionless. Research funds from NASA, the National Science Foundation and the U.S. Department of Energy have been and are being used to build and operate equipment designed to detect the theoretical particles predicted by the Cold Dark Matter theory.

These theoretical particles are called weakly interacting massive particles (WIMPs), some of which are designated neutralinos. Cold dark matter WIMPs were predicted more than 20 years ago, but have never been detected.

In December 2003, Jerome Drexler, the Silicon Valley inventor/scientist, authored a book[5] presenting a contrarian dark matter theory about dark matter protons that orbit small and large galaxies traveling near the speed of light. The book was followed by his 19-page dark matter scientific paper e-Print No. astro-ph/0504512[12] on April 22, 2005. After seeing his dark matter theory/cosmology being supported by a number of astronomical discoveries made by others during 2005, he proceeded to author a 295-page book, *Comprehending and Decoding the Cosmos*[8], that was published on May 22, 2006.

On June 20, 2006, Russia announced[14] that it will launch an ultraviolet astronomical observatory having a 1.7 meter main mirror. The project manager is Boris Shustov, Professor of Physics and Mathematics and head of the Institute of Astronomy at the Russian Academy of Sciences. In the Russian news release he is quoted as saying, "One should particularly emphasize the observatory's role in detecting the so-called dark matter of the universe and unlocking its secrets because such dark matter can only be seen by large ultraviolet telescopes."

"Shustov appears to believe that dark matter particles are charged and extremely fast and therefore not cold," said Drexler, who added that Shustov and he are not alone in concluding that dark matter is not cold.

On Feb. 3, 2006, Professor Gerry Gilmore of Cambridge University and associates announced astronomical data indicating that dark matter is not cold and not collisionless. The 10-page epilogue chapter of Drexler's 2006 book (which now is in Appendix C of this book) reports on Gilmore's dark matter discoveries as follows:

1. Dark matter particles are not slow and cold, but instead appear to be moving at 9 kilometers per second and have an apparent temperature of about 10,000° C, which is higher than the 6,000° C at the surface of the sun.

2. "The strange thing about dark matter is that it doesn't give off radiation."

3. "There must be some form of repulsion [between the dark matter particles]..." "We have to start looking into the physics of the interactions between dark matter particles — not just at the way they respond to gravity."

4. "This indicates that dark matter clumps together in building blocks which have a minimum size," and "This is 1,000 light years across, with 30 million times the mass of the sun." said team member Dr. Mark Wilkinson.

On June 4, 2006 Professor Carl H. Gibson of UC San Diego published a scientific paper, e-Print No. astro-ph/0606073[15] entitled "Cold dark matter cosmology conflicts with fluid mechanics and observations". "According to Gibson's paper, dark matter cannot be cold and collisionless," said Drexler.

Although Sir Martin Rees, The Astronomer Royal of the United Kingdom, has not taken a public stand against Cold Dark Matter, on a web site in 2004[16] he appears to recommend Drexler's December 2003 hot-dark-matter book, among five books recommended in the field of cosmology.

The Dark Matter Scientific Assessment Group (DMSAG), of the Department of Energy and the National Science Foundation, will hold its first meeting on June 29-30, 2006 in Washington, DC. Drexler sent 11 copies of his May 2006 book to DMSAG members. It is hoped that the dark matter news from Russia, Cambridge University, UC San Diego, and Los Altos Hills, California will improve the DMSAG's research strategy, plans and decisions.

CHAPTER 2

COSMIC NEWSWIRE #2

Russian Scientists and Drexler: Dark Matter of the Universe May Radiate Ultraviolet Light

LOS ALTOS HILLS, Calif., July 17, 2006 (AScribe Newswire) -- Ever since the concept of galaxy-orbiting Relativistic-Proton Dark Matter was developed by Jerome Drexler over five years ago, he has been aware that his dark matter model should emit synchrotron radiation of ultraviolet photons, extreme ultraviolet (EUV) photons, or soft X-ray photons. The extragalactic magnetic field is at the right level.

In the past, astronomers doubting Drexler's Relativistic-Proton Dark Matter model had argued that if his model were correct, photons emitted from dark matter halos around spiral galaxies would have been detected by astronomers years ago.

Drexler had responded that the failure of astronomers to detect the synchrotron radiation probably was caused by the low power level of the emitted photons that are generated by relativistic protons moving near the speed of light in circular orbits that form dark matter halos around spiral galaxies. However, protons orbiting groups of galaxies in galaxy clusters are much more energetic and could generate shorter-wavelength photons with hundreds of times more synchrotron radiation power in the extreme ultraviolet (EUV).

Drexler discussed "the far-ultraviolet [wavelength of 1000 angstroms] signature of *missing* baryons" in his December 2003 book *How Dark Matter Created Dark Energy and the Sun.*[5]

The possibility of detection of ultraviolet (or extreme ultraviolet) photons from Relativistic-Proton Dark Matter was reactivated on June 20, 2006, when Russia announced[14] in a news release that it will launch an ultraviolet astronomical observatory having a 1.7 meter main mirror. In the news release, the project manager, Professor Boris Shustov, is quoted as saying, "One should particularly emphasize the observatory's role in detecting the so-called dark matter of the universe and unlocking its secrets because such dark matter can only be seen by large ultraviolet telescopes." This is a very significant newsworthy statement in the fields of both astronomy and astrophysics.

Apparently, Professor Shustov and other Russian scientists believe that dark matter particles are electrically charged, are relativistic in velocity and energy, and generate ultraviolet (or EUV) synchrotron radiation at a power level that can be detected and measured by a large satellite-borne ultraviolet telescope.

This Russian news made a significant contribution supporting Drexler's 2003 Relativistic-Proton Dark Matter model and led to an AScribe news release on this subject on June 27, 2006. That news led a respected professor of physics in the USA, who wishes to remain anonymous, to send emails to Drexler that included the following two quotes supporting Drexler's dark matter model:

> "An extreme ultraviolet and soft X-ray excess has been detected from clusters of galaxies more than ten years ago by EUVE and ROSAT. Today the XMM-Newton satellite continues the research in this exciting field."

> "In this case the EUV and soft X-ray excess from clusters, which is by now a well established phenomenon, could be used to support your [Relativistic-Proton Dark Matter] model."

Over twenty scientific papers have been written on the above subject, sometimes referred as *cluster soft excess*, during the past ten years. A number of these papers point out that the source of the observed EUV excess photon emission from galaxy clusters has not yet been determined or identified, although theories have been suggested.

Drexler posits that the highly energetic galaxy-group-orbiting Relativistic-Proton Dark Matter is probably the leading candidate for such a source of EUV photon emission since its synchrotron radiation provides a very plausible explanation for the *cluster soft excess* phenomenon.

Perhaps the next step to help confirm the Relativistic- Proton Dark Matter model might be to compare astronomical measurements of the extreme ultraviolet excess emission detected from clusters of galaxies to the theoretically derived power spectrum of the radiated EUV photons for the relativistic proton dark matter model for galaxy clusters.

Hopefully, the Dark Matter Scientific Assessment Group of the U.S. Department of Energy and the National Science Foundation will recommend research funding to explore ultraviolet and EUV luminous dark matter.

CHAPTER 3

COSMIC NEWSWIRE #3

Drexler's Dark Matter Relativistic Protons May Be the Cause of Extreme Ultraviolet Emission from Galaxy Clusters

LOS ALTOS HILLS, Calif., August 21, 2006 (AScribe Newswire) -- Extreme ultraviolet (EUV) emission from some spiral galaxy clusters has been observed by astronomers for the past ten years, at much higher levels than the normal EUV emission from the hot gas between the galaxies in clusters. The excess EUV phenomenon has been designated *cluster soft excess* by the astronomers and astrophysicists working in this field.

These researchers say that about 30 percent of the X-ray bright galaxy clusters they have studied exhibit this *cluster soft excess* (CSE) behavior. Thus, the phenomenon represents a double mystery; the mystery of the source of the *cluster soft excess* and the mystery of why only 30 percent of the X-ray bright galaxy clusters exhibit CSE.

Over twenty scientific papers have been written on subjects related to *cluster soft excess*, during the past ten years. A number of these papers point out that the source of the observed EUV excess photon emission from galaxy clusters has not yet been determined or identified, although theories have been suggested. A 2004 book containing a number of these CSE papers is entitled, *Soft X-ray Emission from Clusters of Galaxies and Related Phenomena*[17]. It is edited by Richard Lieu and Jonathan Mittaz of the University of Alabama.

Jerome Drexler recently authored a book, entitled *Comprehending and Decoding the Cosmos*[8], which may provide a possible explanation for the double-mystery CSE phenomenon. One subject of Drexler's May 2006 book is astronomical phenomena involving accelerating dark matter relativistic protons, which also happen to produce extreme ultraviolet photon emission through the well-known astronomical phenomenon known as synchrotron emission (or synchrotron radiation).

For example, synchrotron emission is produced when a relativistic proton, with a velocity near the speed of light, is caused to travel in a circular path by a magnetic field. The synchrotron emission includes ultraviolet photons, extreme ultraviolet (EUV) photons, or soft X-ray photons.

These same three types of photons also can be emitted when a relativistic proton in space is rapidly slowed down by colliding with an atomic nucleus or a dust particle. Since the proton acceleration is much greater in this case than for synchrotron emission, the photon emission probably would be at a shorter wavelength and higher radiation power level. This type of radiation is called Bremsstrahlung, which means *braking radiation* in German.

An observed cluster of spiral galaxies can be in three possible astronomical states: It could be merging with another galaxy cluster, it could have merged with another galaxy cluster in the past, or it could have little or no merger history.

How might the merger history of a cluster of spiral galaxies be determined? It is well known that merging spiral galaxy clusters create starburst galaxies with star formation rates as much as 50 times greater than that for a single spiral galaxy. Also, the new blue and blue-white stars form in the nuclei of starburst spiral galaxies rather than in the spiral arms as in the case of new stars in spiral galaxies. These two features can be used to identify a starburst galaxy.

As explained in Chapters 2006-46 and 2006-47 of Drexler's 2006 book[8], the ultra-high-energy dark matter relativistic protons orbiting groups of galaxies in a cluster that is merging, will be colliding with atoms and dust particles thereby creating the muons that probably catalyze hydrogen fusion nuclear reactions in stars. This collision process also should generate Bremsstrahlung radiation of EUV photons and soft-X-ray photons.

Drexler suggests that astronomers doing CSE research should try to determine whether there is a high correlation between the 30 percent of the galaxy clusters that exhibit the CSE and those galaxy clusters that contain the highest percentage of starburst galaxies. If such a high correlation exists, the explanation for the source of the CSE may be found in Chapters 2006-46 and 2006-47.

A corollary to the propositions of the previous two paragraphs, is that most quiescent observed galaxy clusters with regard to CSE may be clusters that experienced considerable merger and starburst activity in their past and as a result possess a diminished number of orbiting dark matter relativistic protons.

Based upon the above, Drexler posits that dark matter relativistic protons, orbiting groups of galaxies in galaxy clusters, may represent a leading candidate for the source of EUV photon emission since their synchrotron radiation and Bremsstrahlung radiation would provide very plausible explanations for the CSE.

CHAPTER 4

COSMIC NEWSWIRE #4

Dark Matter Was Identified in 2003, Scientifically Confirmed in 2006 in Jerome Drexler's Published Books

LOS ALTOS HILLS, Calif., Aug. 23, 2006 (AScribe Newswire) -- The recent news from NASA regarding the Bullet Cluster[19], confirming the existence of dark matter, was more than welcome to Jerome Drexler. He has spent seven years researching dark matter and has provided overwhelming scientific evidence of the precise identity of the dark matter of the universe.

Although he published two books on his dark matter discoveries, the applause never came, since the existence of dark matter had not yet been confirmed and the 22-year-old unproven Cold Dark Matter theory had not yet been discarded.

In his 2003 book, *How Dark Matter Created Dark Energy and the Sun — An Astrophysics Detective Story*[5], he identifies dark matter and describes the astronomical evidence he found and used to draw that conclusion.

Drexler's new 295-page book discloses the identity of the mysterious dark matter of the universe and its surprising and significant roles and functions in creating spiral galaxies, stars, starburst galaxies, synchrotron radiation, and ultra-high-energy cosmic rays. The paperback book, entitled *Comprehending and Decoding the Cosmos*[8], was published and is sold by Universal Publishers and is also being sold by Amazon.com and Barnes&Noble.com.

Dark matter, the universe's active, massive, extensive, and difficult-to-detect matter, is considered by many to be the greatest mystery of the universe. There are also many other mysteries involving unexplained cosmic phenomena. In his May 2006 sequel, Drexler initially uses 14 of these mysterious phenomena along with his new analytical decoding concept of *dark matter relationism* to discover and identify a promising dark matter candidate compatible with these 14 cosmic phenomena, possibly ending the 70-year quest to identify dark matter.

To test and confirm the validity of his *dark-matter-relationism* based discovery, Drexler finds, analyzes and uses an additional 11 unexplained cosmic phenomena discovered or reported by various astronomers primarily during 2005. Utilizing his same promising dark matter candidate, Drexler explains in a plausible manner all 11 of these recently discovered cosmic mysteries in his May 2006 sequel, further supporting both his dark matter candidate and his new analytical decoding concept of *dark matter relationism.*

As a result, Drexler's *dark matter relationism* research as applied to the 25 unexplained cosmic phenomena has yielded the identification of dark matter and plausible explanations for 15 well-known cosmic mysteries and for 10 lesser-known cosmic mysteries, thereby providing new insights into the nature of the universe and leading to a partial decoding of the cosmos.

Thus, a fitting subtitle for the 2006 book is *Discovering Solutions to Over a Dozen Cosmic Mysteries by Utilizing Dark Matter Relationism, Cosmology, and Astrophysics.*

CHAPTER 5

COSMIC NEWSWIRE #5

Dark Matter's Identity Revealed by Deciphering 14 Cosmic Clues

LOS ALTOS HILLS, Calif., Sept.5, 2006 (AScribe Newswire) -- Dark matter's identity has been discovered through use of a cryptographic-like analysis of 14 constituents of the cosmos.

As a youth during wartime, Jerome Drexler learned how to decipher a 50-word encrypted message or a 50-word encrypted passage from Shakespeare. A decade later in graduate school, a course in Information Theory expanded his knowledge of cryptography.

Drexler has applied a cryptographic-like analysis for solving the mystery of the identity of dark matter (DM) of the universe. Instead of using a 50-word encrypted message to extract the secret code it contains, he used 14 carefully selected cosmic clues called cosmic constituents of the universe to extract the nature and identity of dark matter.

He had speculated that if dark matter represents 80 to 90 percent of the mass of the universe, dark matter should have roles, functions and an influence on most of these 14 cosmic constituents. Each type of dark matter proposed by scientists was subjected to 14 elimination tests as follows.

Drexler asked 14 rhetorical questions: Which type of dark matter (DM) particles could:

1. Form spheroidal dark matter halos around galaxies and DM halos around galaxy clusters?

2. Cause the accelerating expansion of the universe and possibly store dark energy?

3. Be transformed into low velocity hydrogen, protons or proton cosmic rays?

4. Create the magnetic fields within and around spiral galaxies?

5. Be concentrated in the long large curved filaments of dark matter, announced by NASA[18] on September 8, 2004, which form galaxy clusters where two DM filaments intersect (now known as the *cosmic web*)?

6. Create large mature spiral galaxies less than 2.5 billion years after the Big Bang?

7. Create spheroidal DM halos having predictable outer and *hollow* core diameters?

8. Provide angular momentum to spiral galaxies and DM halos?

9. Create galaxies without a central DM density cusp?

10. Create a starless galaxy or a LSB dwarf galaxy with low star formation rates?

11. Lead to linearly rising rotation curves for LSB dwarf galaxies and to flat rotation curves for spiral galaxies?

12. Form 80 percent to 90 percent of the mass of the universe, the remainder being hydrogen, helium, etc?

13. Ignite hydrogen fusion reactions of second-generation stars utilizing hydrogen molecules and dust and ignite fusion reactions of the first generation stars with only hydrogen atoms?

14. Create the first *knee* at 3x1015 eV, the second *knee* between 1017 eV and 1018 eV and the *ankle* at 3x1018 eV of the cosmic-ray energy spectrum near the Earth?

After careful study and analysis, Drexler concluded that galaxy-orbiting relativistic protons would provide many more affirmative answers to the 14 questions than any other known particle. Therefore Relativistic-Proton Dark Matter could be the identity of dark matter since it appears to have the strongest influence on and relationship with the 14 cosmic constituents.

Relativistic-Proton Dark Matter satisfies the three basic requirements of a dark matter candidate. Do such protons have sufficient mass? Yes, relativistic protons can have enormous mass. Have they ever been detected? Yes, relativistic protons bombard Earth's atmosphere every day and are called cosmic rays. Don't relativistic protons move too fast to form small galaxies? The protons can form small galaxies after the protons are slowed down by muon-producing collisions and synchrotron emission losses and after the protons combine with the electrons created by the muon decay, thereby forming hydrogen.

Since protons are electrically charged particles, they would be constrained by the galactic and extragalactic magnetic fields into circular-type orbits forming dark matter halos around galaxies and dark matter around groups of galaxies within galaxy clusters, and also would be concentrated in long large curved filaments of dark matter[18]. All three of these dark matter configurations have been detected by astronomers.

Most of the above information was derived either from Drexler's May 2006 book[8] or his 19-page scientific paper, "Identifying Dark Matter through the Constraints Imposed by Fourteen Astronomically Based Cosmic Constituents", on the Cornell University Library's arXiv.gov website as e-print No. astro-ph/0504512[12].

This astro-ph paper[12] evolved into Drexler's 295-page May 2006 book, *Comprehending and Decoding the Cosmos,* after further research and the inclusion of an additional 11 unexplained cosmic phenomena reported by various astronomers through February 23, 2006. Utilizing the same Relativistic-Proton Dark Matter candidate and the laws of physics, Drexler explains in a plausible manner all 11 of these recently discovered cosmic mysteries, further supporting his Relativistic-Proton Dark Matter candidate.

Drexler's research has led not only to the identification of the dark matter, but also to the discovery of the surprising and significant roles and functions of dark matter in creating spiral galaxies, stars, starburst galaxies, extreme ultraviolet synchrotron radiation, and the ultra-high-energy cosmic rays that bombard Earth. Dark matter appears to be a very active and dynamic medium, not the passive medium represented by Cold Dark Matter.

CHAPTER 6

COSMIC NEWSWIRE #6

Dark Matter Exists! It Solves
25 Cosmic Mysteries

LOS ALTOS HILLS, Calif., Sept. 19, 2006 (AScribe Newswire) -- Jerome Drexler's research in 2005 eventually utilized 25 cosmic mysteries as criteria to uncover the nature and identity of the dark matter of the universe. He was so convinced by his research results he decided to write a sequel to his December 2003 book, which also focused on dark matter.

Thus, while he was authoring *Comprehending and Decoding the Cosmos*[8], during the latter half of 2005, he already was persuaded by his own research of the existence of dark matter.

NASA's August 21, 2006 press release "NASA Finds Direct Proof of Dark Matter"[19] was warmly welcomed by Drexler, since it was sure to diminish the number of dark-matter doubters. That was not to be for very long.

By September 8, two dark-matter-doubting research groups raised issues. The alternative-gravity team at the University of Waterloo in Ontario, Canada, raised one issue and the ether-theory group at Case Western Reserve University in Cleveland, Ohio, raised a possible issue.

The total number of unsolved cosmic mysteries has been increasing year after year. For 2005, in the dark matter field alone, Drexler's 2006 book[8] describes the discovery of 11 new unexplained phenomena. One must conclude that the success of the astronomers is outpacing the ability of the astrophysicists and cosmologists to explain the observed phenomena. (This point is discussed in more detail in Chapters 26, 27, and 28.)

Drexler believes he has explained 25 cosmic mysteries in a plausible manner utilizing one specific form of dark matter and its associated dark matter cosmology. He also believes that the probability of this occurring by chance is very small and highly unlikely.

Drexler invites cosmologists, NASA, the NSF, and the DOE to undertake studies of his research and conclusions, which are already published in his May 22, 2006 book[8].

CHAPTER 7

COSMIC NEWSWIRE #7

In 2008, NASA May Determine if Dark Matter is Cold and Passive or Hot, Active and Ultraviolet Luminous

LOS ALTOS HILLS, Calif., Dec. 5, 2006 (AScribe Newswire) -- On Oct. 31, 2006, NASA announced[20] that it has decided to upgrade the Hubble Space Telescope in the spring of 2008 (now scheduled for August 2008[46]). The new Cosmic Origins Spectrograph (COS) will be installed in the Hubble telescope, increasing its ability to detect ultraviolet (UV) and extreme ultraviolet light (EUV) by a factor of 30 (now estimated at a factor of at least 10).

The COS should make it possible for the upgraded Hubble Space Telescope to detect extreme ultraviolet light if it is emitted at a sufficiently high level from the dark matter within the Milky Way's Local Group galaxy cluster, which also includes the nearby Andromeda galaxy.

Extreme ultraviolet light is very difficult to detect using an Earth-based telescope because the Earth's atmosphere scatters and absorbs a very large percentage of the EUV light passing through it. That is, even if the so-called dark matter of the universe is luminous in the extreme ultraviolet, it still could appear dark through a telescope on Earth.

The possibility of detection of ultraviolet and EUV light from dark matter made the news on June 20, 2006 [14] when Russia announced in a news release that it will launch an ultraviolet astronomical observatory in 2010 having a 1.7 meter main mirror.

CHAPTER 8

COSMIC NEWSWIRE #8

Is Dark Matter a Source of High Energy Gamma Rays? Asks Physical Review Letters; Cosmologist's Books Say 'Yes!'

LOS ALTOS HILLS, Calif., Dec.18, 2006 (AScribe Newswire) -- Earlier this month, Physical Review Letters published a letter titled "H.E.S.S. observations of the Galactic Center region and their possible dark matter interpretation"[21]. One of the authors provided information for a Dec. 8, 2006, article about this research program in PhysOrg.com entitled "Is Dark Matter a Source of High Energy Gamma Rays?"[38].

The HESS gamma-ray observations were made with a High Energy Spectroscopic System (HESS) in the form of an array of four imaging atmospheric Cherenkov telescopes located near the Gamsberg in Namibia.

The PhysOrg.com article ends with the following quote from a HESS researcher, "The probability that we really find dark matter is not that high, but the scientific breakthrough would be tremendous. So it's worth looking for in cosmic gamma rays."

In his 2003 book, *How Dark Matter Created Dark Energy and the Sun[5]*, Jerome Drexler provides significant evidence that the spheroidal dark matter halos around spiral galaxies are comprised of galaxy-group-orbiting high-energy relativistic protons. The 2003 book also provides evidence and references (see pages 48-52) indicating that such relativistic protons could generate gamma rays and x-rays.

Some very relevant gamma-ray-related excerpts from the 2003 book are as follows:

> "Gamma rays are the most powerful form of radiation. They are generated in several ways: Gas at a temperature of 10 billion degrees glows in gamma rays; also, energetic particles smashing into other particles or spiraling through magnetic fields release gamma rays."

> Science News, November 8, 1997, article by Sid Perkins, entitled "Gamma-ray glow bathes Milky Way"[22] states: "A mysterious halo of gamma rays not associated with any known celestial objects extends thousands of light-years from the core of the Milky Way and may surround the entire galaxy, astronomers report."

> "These gamma rays are providing the first evidence that some sort of high-energy process is occurring at large distances from the [Milky Way] galactic core,"

said physicist David D. Dixon of the University of California, Riverside. "The gamma-ray distribution may also provide indirect evidence of dark matter — the universe's missing mass, whose existence scientists have inferred but not yet demonstrated," Dixon said.

Relativistic protons in space are known to produce gamma rays, x-rays, extreme ultraviolet (EUV) and UV emission as these protons are forced to move in circular-type or spiral-type paths by the magnetic fields or when they bombard the giant molecular clouds and dust particles in space. Ultra-high-energy relativistic protons moving in circular or spiral paths generate synchrotron photon emission of UV, EUV or soft x-ray photons. Relativistic protons entering into collisions generate Bremsstrahlung radiation *braking radiation* of x-rays or gamma rays.

Astrophysicists, astronomers, and cosmologists know that these various types of energetic photon emission processes are produced by relativistic cosmic ray protons. However, many of them may not be aware or may not believe that high-energy relativistic cosmic ray protons can be stragglers from dark matter halos. As posited in Drexler's 2003 and 2006 books, some of the dark-matter-halo relativistic protons crossing magnetic fields lose kinetic energy through synchrotron emission or through collisions and thereby become cosmic ray proton stragglers when they are ejected from an orbiting dark-matter proton stream and find their way into a star system such as the sun.

Drexler's books point out that as galaxy orbiting dark matter protons lose kinetic energy and slow down through synchrotron emission losses or particle collisions, the radii of their orbital paths would decline and the protons would penetrate the galaxy as high-energy cosmic ray protons. Some of them would collide with a molecular cloud or with dust before arriving at a star system.

The HESS collaboration team also published an earlier paper in Nature[23] in February 2006 titled "Discovery of Very-High-Energy Gamma-Rays from the Galactic Centre Ridge." It states, "The hardness of the Gamma-ray spectrum and the conditions in those molecular clouds indicate that the cosmic rays giving rise to the Gamma-rays are likely to be protons and nuclei rather than electrons." This statement by the HESS researchers supports Drexler's relativistic proton dark matter hypothesis since the researchers/astronomers confirm that relativistic protons, bombarding molecular clouds, generate the observed gamma rays and they rule out energetic electrons as the source of energy for the observed gamma rays.

Drexler's 2006 astro-cosmology book[8], a sequel to his 2003 dark-matter book[5], is titled *Comprehending and Decoding the Cosmos: Discovering Solutions to Over a Dozen Cosmic Mysteries by Utilizing Dark Matter Relationism, Cosmology and Astrophysics.*

CHAPTER 9

COSMIC NEWSWIRE #9

So-Called Anomalies in NASA-Hubble 3D Dark Matter Map Are Explained by Astro-Cosmology Author Jerome Drexler

LOS ALTOS HILLS, Calif., Jan.16, 2007 (AScribe Newswire) -- The recent 3D mapping of a huge ring of dark matter in the galaxy cluster Cl0024+17 [In May 2007, a paper was published as arXiv: 0705.2171,[24] Title: Discovery of a Ringlike Dark Matter Structure in the Core of the Galaxy Cluster Cl0024+17], is a major astronomical accomplishment of NASA and the Hubble Space Telescope. However, the researchers' reports of so-called discrepancies and anomalies in the distribution of dark matter relative to the distribution of ordinary matter has prevented the 3D mapping from being an immediate cosmological success.

A study of the researchers' Jan. 7 comments during a press conference following the presentation of their scientific paper at the 209th meeting of the American Astronomical Society (AAS) in Seattle, Wash., provides some insight into the reasons behind the researchers' reports of apparent discrepancies and anomalies. (See their comments below.)

Based upon their stated concerns about the so-called discrepancies and anomalies, the researchers apparently believe in the bottom-up theory of galaxy formation, not the top-down theory of galaxy formation, which has some support including strong support of Jerome Drexler.

The astronomical data the researchers are concerned about actually support the top-down theory of galaxy formation and if the data had been described in that manner there would not have been the issue of discrepancies or anomalies — and the 3D mapping might have been considered a cosmological success as well as an astronomical success.

Some relevant comments by some of the dark-matter-map researchers at an AAS press conference in Seattle[36] on Jan. 7, 2007 are as follows:

"We have to resolve discrepancies in the otherwise strong connection between ordinary matter and dark matter."

"Finding what I would call *naked* clumps of dark matter where there are no galaxies for me is very strange. All dark matter clumps of sufficient size should have galaxies — if our understanding is correct."

"We see that dark matter concentrations sometimes seem to have no corresponding ordinary matter."

"For the moment, no one is talking about needing to revise cosmological models; but everything hinged on the size of these anomalies."

"The discrepancies could turn out simply to be artifacts, caused by noise in the data. But then again, they could be real."

"A researcher said the anomalies were *tantalizing* and that his team was eager to investigate them more closely."

The bottom-up theory, where small galaxies form first and then merge over time to form large galaxies, is known to have serious flaws. For example, note that at the same AAS conference, New York University researchers gave a paper based upon their astrophysics paper titled "A New Force in the Dark Sector?"[25]. Their paper states, "The number of superclusters observed in SDSS data appears to be an order of magnitude (about ten times) larger than predicted by Lambda-Cold Dark Matter simulations."

Since Lambda-Cold Dark Matter simulations are based on the bottom-up theory of galaxy formation, the NYU researchers are indicating that the bottom-up theory is extremely inaccurate in predicting the number of superclusters in the universe. Therefore the so-called anomalies and discrepancies reported by the 3D dark-matter-map researchers may have evolved simply because the researchers tested the 3D astronomical data against the wrong galaxy-formation theory. The NASA-Hubble researchers should now test the so-called discrepancies and anomalies against the top-down theory of galaxy formation. (They may discover that Drexler's top-down theory will predict a higher number of superclusters in the universe.)

In the top-down theory of galaxy formation the well-known long, large, slightly curved dark matter filaments (now considered part of the *cosmic web*) form potential galaxy clusters where such dark matter filaments intersect and collide[18,47]. Then small and large galaxies form in this galaxy cluster region from the remnants of the collisions of the intersecting dark matter filaments.

The top-down theory of galaxy formation is further explained in the pages of ten Index references in Jerome Drexler's May 2006, 295-page astro-cosmology book titled *Comprehending and Decoding the Cosmos: Discovering Solutions to Over a Dozen Cosmic Mysteries by Utilizing Dark Matter Relationism, Cosmology, and Astrophysics*[8].

CHAPTER 10

COSMIC NEWSWIRE #10

Relativistic Dark Matter May Solve Big Bang Enigma, Says Astro-Cosmology Author Jerome Drexler

LOS ALTOS HILLS, Calif., Jan. 29, 2007 (AScribe Newswire) -- In a surprising manner, the Big Bang may have satisfied the Second Law of Thermodynamics. An understanding of this phenomenon is helped by an excerpt from Stephen Hawking's earlier tutorial on the subjects of disorder, entropy, the Second Law of Thermodynamics, and the arrow of time[10]:

"It is a matter of common experience that things get more disordered and chaotic with time. This observation can be elevated to the status of a law, the so-called Second Law of Thermodynamics. This says that the total amount of disorder, or entropy, in the universe, always increases with time."

If the amount of disorder, or entropy, in the universe always increases with time, then at the beginning of time the entropy must have been at its lowest level. The Big Bang also occurred at the beginning of time. Therefore, if we accept the Second Law of Thermodynamics, we must also accept that immediately after the Big Bang the entropy of the universe would be at the lowest level it would reach throughout all time.

However, the Big Bang is normally characterized as a chaotic massive fireball explosion associated with a high level of disorder and entropy. We are thus faced with an enigma as to the level of entropy following the Big Bang, but we are not alone.

On November 18, 2004, the University of Chicago published an article entitled "Astrophysicists attempt to answer the mystery of entropy"[11] that contains the following relevant two-sentence paragraph: "But the mystery remains as to why entropy was low in the universe to begin with. The difficulty of that question has long bothered scientists, who most often simply leave it as a puzzle to answer in the future."

If the entropy following the Big Bang had been very low the Second Law of Thermodynamics would have been satisfied, but how could a fiery, chaotic fireball Big Bang explosion have low entropy? This is the enigma that "has long bothered scientists."

Jerome Drexler sees a possible solution to this enigma that would have the Big Bang firing out, in all directions, high-speed ultra-high-energy (UHE) relativistic protons and helium nuclei in a nuclei ratio of about 12:1; In other words, *a violent radial dispersion of relativistic baryons.*

A very high percentage of the energies of these relativistic nuclei would be available to do work in the universe while their entropy, the measure of the amount of their energy which is unavailable to do work, would be very low. Such a Big Bang, characterized by a *violent radial dispersion* of UHE relativistic nuclei, could create an ultra-high usable energy and an ultra-low entropy, and could be designated a *Relativistic Big Bang.*

The temperature of a *Relativistic Big Bang* could be estimated by averaging the kinetic energies of the relativistic protons and helium nuclei. The estimated temperature would be extremely high and probably of the same order of magnitude as the temperature scientists estimate for the Big Bang. Nevertheless, the *Relativistic Big Bang* would have the very low entropy that the Second Law of Thermodynamics requires for the beginning of time.

Some astronomical evidence for a *Relativistic Big Bang* (RBB) comes from the ultra-high-energy cosmic ray (UHECR) protons that bombard Earth's atmosphere every day. The RBB is the most plausible origin of the UHECR's. In Drexler's Relativistic-Proton Dark Matter theory these UHECR's are stragglers from the galaxy-orbiting UHE relativistic protons that form the dark matter streams in the halos surrounding galaxies and groups of galaxies.

It is widely accepted that the mass of dark matter today totals about 83 percent of the mass of the universe and that dark matter was created by the Big Bang. (Drexler's top-down theory of galaxy formation puts this percentage closer to 100 percent during the Big Bang, then during the following 13.7 billion years an estimated 10 to 15 percent of the nuclei attracted electrons and lost kinetic energy and became hydrogen and helium.) Because of this very strong Big Bang-dark matter linkage, strong evidence of the existence of Relativistic-Proton Dark Matter would provide strong evidence for the existence of the *Relativistic Big Bang*. Drexler believes that his 2003[5] and 2006[8] books, his 2005[12] scientific paper, and his 2006/2007 scientific newswires provide very strong scientific evidence for the existence of Relativistic-Proton Dark Matter and therefore for the existence of the RBB.

Cosmological support for a RBB could come via compatibility with, for example, the CMB, or Cosmic Inflation (see Chapters 31 and 34), or the Second Law of Thermodynamics, or the temperatures of the Big Bang, or entropy magnitudes, or the mass values estimated for dark matter particles, or the fact that 83 percent of the universe mass is dark matter. Note that a RBB would be a very efficient way of creating a universe and conserving its energy because the fewest number of particles and the most useful energy would be created and dispersed. These characteristics appear to be especially compatible with Cosmic Inflation theory and its associated Big Bang.

The title of Drexler's December 2003 book is, *How Dark Matter Created Dark Energy and the Sun.* The title of his April 22, 2005, 19-page scientific paper[12] is "Identifying Dark Matter through the Constraints Imposed by Fourteen Astronomically Based 'Cosmic Constituents'".

The title of Drexler's May 2006 book is *Comprehending and Decoding the Cosmos: Discovering Solutions to Over a Dozen Cosmic Mysteries by Utilizing Dark Matter Relationism, Cosmology, and Astrophysics*[8].

This 2006 book provides strong scientific evidence that the dark matter of the universe is comprised of relativistic protons orbiting galaxies and groups of galaxies. This is demonstrated in the book by utilizing the Relativistic-Proton Dark Matter hypothesis, in conjunction with the laws of physics, to derive explanations for more than 15 unsolved cosmic mysteries.

CHAPTER 11

COSMIC NEWSWIRE #11

Astronomers Must Maximize Knowledge Derived From GLAST and UV-COS/Hubble Astronomical Data, Says Astro-Cosmology Author Jerome Drexler

LOS ALTOS HILLS, Calif., Feb. 5, 2007 (AScribe Newswire) -- The total number of unsolved cosmic mysteries has been increasing year after year. For 2005, in the dark matter field alone, Drexler's 2006 book[8] describes the discovery of 11 new unexplained phenomena. One must conclude that the success of the astronomers is outpacing the ability of the astrophysicists and cosmologists to explain the observed phenomena. (This point is discussed in more detail in Chapters 26, 27, and 28.)

Also, astronomers are building more and more advanced telescopic systems by utilizing space platforms, employing adaptive optics, and by combining images derived from photons of different wavelengths. With the cost of such projects sky-rocketing and Russia catching up, US astronomers must be trained to maximize the knowledge derived from astronomical data. Examples in 2008:

GLAST's gamma rays and the COS/Hubble's ultraviolet rays.

These goals can be achieved by establishing and utilizing new graduate courses, new textbooks, seminars, case study methods, or visiting lecturers, to cover relevant academic subjects. The subjects might include information theory, statistical analysis, cryptography, the interpretation of astronomical data, current limitations of telescopic methods, Occam's razor logic, astrophysical relationism, recent cosmology history, theories of galaxy formation, and dark matter cosmology. Let us look at three examples where this proposed broader education might have been or can be very helpful.

Today, most dark matter searches are being done in deep underground mines in order to shield the detectors and instrumentation from being bombarded by ultra-high-energy cosmic ray protons. But Jerome Drexler has posited that these same cosmic ray protons being shielded from the researcher's underground detectors are the dark matter particles for which they have been searching for years.

Maybe this underground effort will end soon. On Oct. 31, 2006, NASA announced[20,46] that it will upgrade the Hubble Space Telescope in the spring [now August] of 2008. The new Cosmic Origins Spectrograph (COS) will be installed in the Hubble, increasing its ability to detect UV and extreme

UV by a factor of 30 (now 10). Also, Russia announced in June 2006 that dark matter particles emit UV photons and that Russia will launch a space-based EUV/UV telescope in 2010 to detect them[14].

In a scientific paper[26] on the use of the Gamma-ray Large Area Space Telescope (GLAST) to detect particle dark matter, an assumption was made that gamma-rays are produced by pair annihilation of dark matter Weakly Interacting Massive Particles (WIMPS). Since WIMPs have not been detected after over 20 years of searches, they may not exist.

Drexler believes that dark matter comprises relativistic protons, orbiting galaxy groups that emit gamma rays when the protons enter into collisions with dust and molecules and generate Bremsstrahlung radiation. Since GLAST will be a space-based telescope with limited future access, he feels that GLAST should be modified to enable it to detect both proton-generated gamma rays or WIMP-annihilation generated gamma rays. The GLAST mission is scheduled to launch in the fall of 2007 (now in 2008) from the Kennedy Space Center.

The recent 3D mapping of a ring of dark matter is a major astronomical accomplishment of NASA and the Hubble telescope. However, the astronomers' reports of so-called discrepancies and anomalies in the distribution of dark matter relative to the distribution of ordinary matter has prevented the 3D mapping from being an immediate cosmological success.

However, based upon the astronomers' stated concerns about the so-called anomalies and discrepancies, they apparently believe in the bottom-up theory of galaxy formation, not the top-down theory. The astronomical data the astronomers were concerned about appear to support the top-down theory of galaxy formation and if these data had been described to the public in that manner there would not have been an issue of *discrepancies* or *anomalies*.

Drexler discovered and developed both the concept of *dark matter cosmology* and methods to maximize the amount of knowledge that can be derived from astronomical data. He also searched for and found a number of scientific papers that reported astronomical data in conjunction with a remaining cosmological mystery or enigma. He used this substantial amount of astronomical data, ideas from his December 2003 book[5], and his analytical methods to derive, in his 2006 book[8], plausible explanations for at least 15 and up to as many as 25 cosmic mysteries.

CHAPTER 12

A Relativistic-Proton Dark Matter Would Be Evidence the Big Bang Probably Satisfied the Second Law of Thermodynamics*

Jerome Drexler
Former NJIT Research Professor of Physics
New Jersey Institute of Technology

*This is derived from Drexler's Feb. 15, 2007 scientific paper
available at   http://arxiv.org/abs/physics/0702132.

Abstract

A new research hypothesis has been developed by the author based upon finding astronomically based *cosmic constituents* of the universe that may be created or influenced by or have a special relationship with possible dark matter (DM) candidates. He then developed a list of 14 relevant and plausible *cosmic constituents* of the Universe, which then was used to establish a list of constraints regarding the nature and characteristics of the long-sought dark matter particles. A dark matter candidate was then found that best conformed to the 14 constraints established by the *cosmic constituents*. The author then used this same dark matter candidate to provide evidence that the Big Bang could be characterized as a *violent radial dispersion of relativistic baryons*, had a low entropy, and therefore probably satisfied the Second Law of Thermodynamics.

Determining the Nature of the Dark Matter of the Universe

One hundred years ago, Albert Einstein announced the Special Theory of Relativity, which predicted and explained that a proton traveling near the speed of light could have a relativistic mass a thousand, a million, or even a billion times greater than the mass of a proton at rest. (This led the author to conceive his dark matter theory. The idea occurred to him that the gravitational strength of multitudinous galaxy-orbiting relativistic protons moving in the cosmos could create extremely large gravity-related tidal forces on nearby matter, like that exhibited by dark matter.)

Astronomer Fritz Zwicky[1] discovered the presence of dark matter in the Coma cluster of galaxies in 1933. Ever since astronomer Vera Rubin[2,3] confirmed the existence of dark matter halos around galaxies in 1977, cosmologists and astrophysicists have been trying to identify the dark matter particles.

In 1984, scientists[4] developed a Cold Dark Matter (CDM) theory based upon a theoretical uncharged, slow moving particle that they called the Weakly Interacting Massive Particle (WIMP). More recently, it was estimated by scientists that the theoretical WIMP dark matter particles would require a mass in the range of about 35 to 10,000 times[5] greater than the mass of a proton at rest in order to exhibit the observed gravity-related forces of dark matter halos. However, searches for the theoretical WIMP particles during the past 20 years have all come up empty handed.

For this reason, and knowing that Einstein's relativistic proton easily could meet the mass requirement of the mysterious dark matter particles and that relativistic cosmic ray protons are widely observed, the author has endeavored to determine the nature of dark matter.

The author posits that relativistic protons, orbiting galaxies, have the necessary characteristics of the long-sought dark matter particles, which are estimated by most scientists to comprise 80% to 90%[6] of the total mass of the universe. Relativistic protons do have the required mass and the required difficulty of detection. Protons also can transform themselves into hydrogen, the principal matter of galaxies, by creating muons[7,8] that decay into electrons, then combining with the electrons.

Thus, relativistic protons could form (1) galaxies and their dark matter halos, (2) galaxy clusters and their internal dark matter, and (3) the long, large, filamentary dark matter known[18,47] to crisscross the cosmos [now called the cosmic web].

However, for this proton-based dark matter theory to become widely accepted, there also should be astronomical evidence of relativistic protons within the dark matter halo surrounding the Milky Way. The author posits that the high-energy cosmic ray relativistic protons bombarding Earth's atmosphere every day, from all directions, lend credence toward providing such astronomical evidence.

The author has applied a cryptographic-like analysis for solving the mystery of the identity of dark matter of the universe. Instead of using an encrypted message to extract the secret code it contains as in normal cryptography, the author used 14 cosmic constituents of the universe to extract the nature and identity of dark matter.

The author had speculated that if dark matter represents 80% to 90% of the mass of the universe, dark matter should have roles, functions or an influence on most of the following 14 cosmic constituents. Each type of dark matter proposed by scientists was subjected to 14 elimination tests as follows.

The author asked 14 rhetorical questions: Which type of dark matter (DM) particles could:

1. Form spheroidal dark matter halos around galaxies and DM halos around galaxy clusters?

2. Cause the accelerating expansion of the universe and possibly store dark energy?

3. Be transformed into low-velocity hydrogen, protons, or proton cosmic rays?

4. Create the magnetic fields within and around spiral galaxies?

5. Be concentrated in the long, large, curved filaments of dark matter announced by NASA on September 8, 2004[18,47] [now called the *cosmic web*], which form galaxy clusters where two DM filaments intersect?

6. Create large, mature, spiral galaxies less than 2.5 billion years after the Big Bang?

7. Create spheroidal DM halos having predictable outer and *hollow* core diameters?

8. Provide angular momentum to spiral galaxies and DM halos?

9. Create galaxies without a central DM density cusp?

10. Create a starless galaxy or a Low Surface Brightness (LSB) dwarf galaxy with low star formation rates?

11. Lead to linearly rising rotation curves for LSB dwarf galaxies and to flat rotation curves for spiral galaxies?

12. Form 80% to 90% of the mass of the universe, the remainder being hydrogen, helium, etc.?

13. Ignite hydrogen fusion reactions of second generation stars utilizing hydrogen molecules and dust and ignite fusion reactions of the first generation stars with only hydrogen atoms?

14. Create the first *knee* at 3×10^{15} eV, the second *knee* between 10^{17} eV and 10^{18} eV, and the *ankle* at 3×10^{18} eV of the cosmic-ray energy distribution at the Earth? (See Appendix B, Slide #17)

After careful study and analysis, the author concluded that galaxy-orbiting relativistic protons would provide many more affirmative answers to the 14 questions than any other known particle. Therefore, Relativistic-Proton Dark Matter could be the identity of dark matter since it appears to have the strongest influence on and relationship with the 14 *cosmic constituents*. This dark matter identification procedure could also be described as utilizing Ockham's (Occam's) razor logic 14 times.

Relativistic-Proton Dark Matter satisfies the three basic requirements of a dark matter candidate. Do such protons have sufficient mass? Yes, relativistic protons can have enormous mass. Have they ever been detected? Yes, relativistic protons bombard Earth's atmosphere every day and are called cosmic rays. Don't relativistic protons move too fast to form small galaxies? The protons can form small galaxies after the protons are slowed down by muon-producing[7,8] collisions and synchrotron emission energy losses, and after the protons combine with the electrons created by the muon decay, thereby forming hydrogen.

Since protons are electrically charged particles, they would be constrained by the weak extragalactic and galactic magnetic fields into extremely large circular/spiral orbits forming dark matter halos around galaxies and dark matter around groups of galaxies within galaxy clusters, and also could be concentrated into long large curved filaments of dark matter. All three of these dark matter configurations have been reported by astronomers.

Much of the above information was derived from the author's May 2006 book[8] and his 19-page April 2005 paper, "Identifying Dark Matter through the Constraints Imposed by Fourteen Astronomically Based Cosmic Constituents"[12], found on the arXiv.gov website as e-print No. astro-ph/0504512.

The author's 295-page May 2006 book, *Comprehending and Decoding the Cosmos*, analyzes an additional 11 cosmic enigmas beyond the 14 derived from his astro-ph paper[12]. Utilizing only Relativistic-Proton Dark Matter theory and the laws of physics, the author explains in a plausible manner all 11 of these recently discovered cosmic enigmas, further supporting the Relativistic-Proton Dark Matter theory.

The author's research has led not only to the identification of the dark matter but also to the discovery of the surprising and significant roles and functions of dark matter in creating the cosmic web, spiral galaxies, stars, starburst galaxies, extreme ultraviolet synchrotron radiation, and the ultra-high-energy cosmic rays that bombard Earth.

Dark matter appears to be a very active and dynamic medium comprising relativistic protons and helium nuclei in the well-known ratio of about 12: 1. Dark matter is widely believed to represent 80% to 90% of the mass of the universe, and believed to be created by the Big Bang. These dark matter characteristics provide the evidence required and used in the next section to reach the conclusion that the Big Bang was relativistic, had a low entropy, and probably satisfied the Second Law of Thermodynamics.

A Relativistic-Proton Dark Matter Would Be Evidence that the Big Bang Had Low Entropy and Probably Satisfied The Second Law of Thermodynamics

In a surprising manner, the Big Bang may have satisfied the Second Law of Thermodynamics. An understanding of this phenomenon is helped by an excerpt from Stephen Hawking's earlier tutorial[10] on the subjects of disorder, entropy, the Second Law of Thermodynamics, and the arrow of time: "It is a matter of common experience, that things get more disordered and chaotic with time. This observation can be elevated to the status of a law, the so-called Second Law of Thermodynamics. This says that the total amount of disorder, or entropy, in the universe, always increases with time."

If the amount of disorder, or entropy, in the universe always increases with time, then at the beginning of time the entropy must have been at its lowest level. The Big Bang also occurred at the beginning of time. Therefore, if we accept the Second Law of Thermodynamics, we must also accept that immediately after the Big Bang the entropy of the universe would be at the lowest level it would reach throughout all time.

However, the Big Bang is normally characterized as a fiery, chaotic, fireball explosion associated with a high level of disorder and entropy. We are thus faced with an enigma as to the level of entropy following the Big Bang, but we are not alone.

On November 18, 2004, the University of Chicago published an article[11] entitled "Astrophysicists attempt to answer the mystery of entropy" that contains the following relevant two-sentence paragraph: "But the mystery remains as to why entropy was low in the universe to begin with. The difficulty of that question has long bothered scientists, who most often simply leave it as a puzzle to answer in the future."

If the entropy following the Big Bang had been very low, the Second Law of Thermodynamics would have been satisfied, but how could a fiery, chaotic Big Bang explosion have low entropy? This is the enigma that "has long bothered scientists."

The author sees a possible solution to this enigma that would have the Big Bang firing out, in all directions, high-speed ultra-high-energy (UHE) relativistic protons and helium nuclei near the well-known atomic ratio of 12:1; In other words, a *violent radial dispersion of relativistic baryons.*

A very high percentage of their energies would be available to do work in the universe while their entropy, the measure of the amount of their energy which is unavailable to do work, would be very low. Such a Big Bang, characterized by a *violent radial dispersion* of UHE relativistic nuclei, could create very high usable energy and very low entropy, and could be designated a *Relativistic Big Bang* (RBB).

The temperature of a *Relativistic Big Bang* could be estimated by averaging the kinetic energies of the relativistic protons and helium nuclei. The estimated temperature probably would be of the same order of magnitude as the temperature that scientists estimate for the Big Bang. Nevertheless, the *Relativistic Big Bang* would have the very low entropy that the Second Law of Thermodynamics requires for the *beginning of time.*

Some astronomical evidence for a *Relativistic Big Bang* comes from the ultra-high-energy cosmic ray (UHECR) protons that bombard the Earth's atmosphere every day. The RBB is the most plausible origin of these UHECRs (See Chapters 29, 31, and Appendix A regarding the GZK effect.)

In the author's Relativistic-Proton Dark Matter theory, these UHECRs are stragglers from the UHE relativistic protons that orbit groups of galaxies within galaxy clusters.

It is widely accepted that the mass of dark matter today totals about 83% of the mass of the universe and that dark matter was created by the Big Bang. Because of this very strong Big Bang-dark matter linkage, strong evidence of the existence of relativistic-proton dark matter would provide strong evidence for the existence of the *Relativistic Big Bang*. The author believes that his 2003[5] and 2006[8] books and his 2005[12] scientific paper provide very strong scientific evidence for the existence of Relativistic-Proton Dark Matter and, therefore, for the existence of the RBB.

Cosmological support for an RBB may come eventually via compatibility with, for example, the Cosmic Microwave Background, or Cosmic Inflation (see Chapters 31 and 34), or the Second Law of Thermodynamics, or the temperatures of the Big Bang, or the mass values for dark matter particles, or the 83% dark matter mass. Note that an RBB would be a very efficient way of creating the universe and conserving its energy because the fewest number of particles and the most useful energy would be created and dispersed, which are characteristics that may be compatible with Cosmic Inflation theory (see Chapters 31 and 34) and its associated Big Bang.

As previously indicated, strong scientific evidence that the dark matter of the universe is comprised of galaxy-orbiting relativistic protons can be found in the 2003 book *How Dark Matter Created Dark Energy and the Sun*[5], the 2005, 19-page scientific paper "Identifying Dark Matter Through the Constraints Imposed by Fourteen Astronomically Based 'Cosmic Constituents'"[12], and the 2006 book, *Comprehending and Decoding the Cosmos: Discovering Solutions to Over a Dozen Cosmic Mysteries by Utilizing Dark Matter Relationism, Cosmology, and Astrophysics* [8].

Confirmation of the identification of dark matter is scientifically supported in the 2006 book through the utilization of the Relativistic-Proton Dark Matter hypothesis, in conjunction with the laws of physics, to derive solutions and plausible explanations for more than 15 previously unsolved cosmic mysteries.

If the existence of the Relativistic-Proton Dark Matter provides strong evidence that the Big Bang satisfied the Second Law of Thermodynamics, then a corollary could follow: Since the Big Bang must have satisfied the Second Law of Thermodynamics, its entropy must have been very low; and since relativistic protons possess the highest possible energy and the lowest possible entropy, they must have represented the principal mass output of the Big Bang.

References for this Paper

1. F. Zwicky, 1937 *Astrophys. J.* (Lett) 86, 217

2. V. C. Rubin, N. Thonnard and W. K.Ford, 1978 *Astrophys. J.* (Lett) 225, L107

3. V. Rubin, *Bright Galaxies – Dark Matters* (Amer. Inst. Physics, New York, 1997), p. 109-116

4. G. Blumenthal, S. Faber, J. R. Primack, and M. J. Rees, 1984 *Nature* **311**, 517

5. J. Drexler, *How Dark Matter Created Dark Energy and the Sun* (Universal Publishers, Parkland, Florida, USA, 2003), p. 16

6. J. Drexler, *op. cit.*, p. 18

7. A. S. Bishop, *Project Sherwood – The U.S. Program in Controlled Fusion* (Addison- Wesley Publishing Company, Inc., Reading, Massachusetts, U.S.A. 1958) p 177-178

8. J. Drexler, *Comprehending and Decoding the Cosmos* (Universal Publishers, Boca Raton, Florida, USA, 2006)

9. J. Drexler, 2005, astro-ph/0504512 v1

10. S. Hawking, Lecture – Life in the Universe, http://www.hawking.org.uk/lectures/life.html

11. S. Koppes, The University of Chicago Chronicle, Vol. 24, No.5, November 18, 2004, http://chronicle.uchicago.edu/041118/entropy.shtml

CHAPTER 13

COSMIC NEWSWIRE #12

Big Bang Was Not a Chaotic Fireball Explosion, But a Violent Radial Dispersion of Relativistic Baryons That Became Dark Matter, New Scientific Paper Posits

LOS ALTOS HILLS, Calif., Feb. 19, 2007 (AScribe Newswire) -- A new scientific paper published and available on the Internet[13], posits that the Big Bang was not a fiery, chaotic, disordered, fireball explosion but a *violent radial dispersion* of ultra-high velocity dispersion of ultra-high-energy relativistic protons and helium nuclei having a nuclei ratio in the range of about 12:1.

The paper explains that the Big-Bang dispersed relativistic protons and helium nuclei evolved into the mysterious dark matter that now represents about 83 percent of the mass of the universe. In other words, a Big Bang in the form of a *violent radial dispersion of relativistic baryons* created the dark matter of the universe.

The paper, published on the Cornell University Library arXiv.org physics website, is entitled "A Relativistic-

Proton Dark Matter Would Be Evidence The Big Bang Probably Satisfied The Second Law of Thermodynamics". It is dated February 15, 2007 and is available at: http://arxiv.org/abs/physics/0702132.

One of the various types of astronomical evidence supporting such a relativistic Big Bang is the ultra-high-energy cosmic ray (UHECR) protons that bombard Earth's atmosphere every day. In Jerome Drexler's Relativistic-Proton Dark Matter theory, these UHECR's are considered to be stragglers from the UHE relativistic protons that orbit groups of galaxies as dark matter particles in galaxy clusters.

It is widely accepted that dark matter represents 83 percent of the output of the Big Bang. Because of this strong Big Bang-dark matter linkage, the strong evidence of the existence of Relativistic-Proton Dark Matter provided in Drexler's 2003[5] and 2006[8] books also provides strong evidence for the existence of a Relativistic Big Bang.

CHAPTER 14

COSMIC NEWSWIRE #13

Dr. Stephen Petrina's 'Change and Technology in the U.S.' Credits Silicon Valley's Jerome Drexler for Invention of the Laser Optical Storage System

LOS ALTOS HILLS, Calif., March 5, 2007 (AScribe Newswire) -- Earlier in his career, dark matter cosmologist Jerome Drexler pioneered the laser recording of data using his co-invention, in 1979, of Drexon®, a sealed nanotechnology-based laser recording media.

In 1981 Jerome Drexler and Drexler Technology Corporation of Silicon Valley (now called LaserCard Corporation, Nasdaq: LCRD), won the IR 100 Industrial Research Award for the world's first laser read/writable optical memory disc for digital information storage.

Dr. Stephen Petrina, Associate Professor at the University of British Columbia, now gives Jerome Drexler sole credit for the invention and technological development of the Laser Optical Storage System. That designation is on page 89 of Dr.Petrina's recently published history of technology,

Change and Technology in the United States: A Resource Book for Studying the Geography and History of Technology.

Drexler also invented the LaserCard® optical memory card in 1981. More than 35 million Drexon-based LaserCard optical memory cards have been sold to date. Currently active major optical memory card programs include the US Green Card, the US Laser Visa Mexican-border-crossing card, the US DOD logistics card, the Canadian permanent resident card, the Italian national ID card, the Italian permanent resident card, vehicle registration cards of several states of India, and the Saudi Arabian National ID card.

Jerome Drexler and Eric W. Bouldin, then of Drexler Technology Corp., jointly filed their first Drexon patent application on 7/6/79 that became USP 4,278,756. They then filed patent applications in 1979-1980 that led to US patents 4,269,917; 4,284,716; and 4,298.684. The patent claims are limited to structures formed from silver particles with maximum dimensions of 50 nanometers, to ensure low-laser-power recording.

In 1986 the term *nanotechnology* entered the English language via the first book on that subject. Later, nanotechnology became defined as structures utilizing building block particles "in the length scale of approximately 1 to 100 nanometer range." Thus, Jerome Drexler and Eric

Bouldin invented a commercially successful nanotechnology product seven years before the first publication of the word *nanotechnology*.

These days, Drexler is deeply involved in astro-cosmology research. In 2002 he discovered and developed the concept of *dark matter cosmology*, based on Albert Einstein's 1905 Special Theory of Relativity. He has authored books published in 2003[5] and 2006[8] and scientific papers in 2005 (astro-ph/0504512)[12] and in 2007 (physics/0702132)[13] on the physics arXiv, to explain and to provide scientific support for his theory of *dark matter cosmology*.

In recent years, Drexler further developed both the concept of *dark matter cosmology* and methods to maximize the amount of knowledge that can be derived from astronomical data. He used a substantial amount of astronomical data, ideas from his 2003 book, and his new analytical methods to derive in his 2006 book plausible explanations for at least 15 and up to as many as 25 mysteries of the cosmos.

The title of Drexler's December 2003 book is *How Dark Matter Created Dark Energy and the Sun: An Astrophysics Detective story*[5].

The title of Drexler's May 2006 book is *Comprehending and Decoding the Cosmos: Discovering Solutions to Over a*

Dozen Cosmic Mysteries by Utilizing Dark Matter Relationism, Cosmology, and Astrophysics[8].

CHAPTER 15

COSMIC NEWSWIRE #14

Big Bang Discovery: Cold Dark Matter May Not Exist, but Einstein-Based Hot Dark Matter Should

LOS ALTOS HILLS, Calif., March 14, 2007 (AScribe Newswire) -- A four-page article in The New York Times Magazine entitled, "Out There"[43], by Richard Panek on Sunday, March 11, focused on the extreme pessimism that prevails today among the many dark matter physicists who have been searching for the mysterious Cold Dark Matter of the universe for as long as 15 years, without success.

The article did not mention Silicon Valley's inventor/ scientist, Jerome Drexler, who entered the race to identify dark matter in 2002, based upon Albert Einstein's 1905 Special Theory of Relativity. He has authored books published in 2006[8] and 2003[5] and two scientific papers[12,13] on his Einstein-based *hot* dark matter theory. As described in the 2006 book, his theory appears to explain 15 to 25 previously *unsolved* cosmic-phenomena mysteries.

This May 2006 book is available now at libraries of 22 (now over 40) prominent universities and astronomical institutes including: Harvard, Harvard-Smithsonian, Yale, Stanford, UC Berkeley, Cornell, Vassar, University of Toronto, University of Edinburgh , University of Helsinki, Kyoto University, Universidad de Chile, University of Hamburg, University of Bologna, University of Goettingen, Canterbury University, Max-Planck-Institut for Astrophysik, Pontificia Universidad Catolica de Chile, Astronomical Institute of the Russian Academy of Sciences, University of Groningen, Universidad de Guadalajara, and the Czech Republic's Academy of Sciences.

Drexler's discovery of a strong linkage between the dark matter of the universe and the nature of the Big Bang, indicates that both the Big Bang and the dark matter it created must have satisfied the Second Law of Thermodynamics. Cold Dark Matter could not have done so since the Second Law (and probably the First Law) would have been violated in the transformation of the output of an enormous-temperature Big Bang immediately into Cold Dark Matter.

Drexler's recent scientific paper[13], published and available on the Internet, posits that the Big Bang was not a fiery, chaotic, disordered, fireball explosion but an orderly ultra-high velocity dispersion of relativistic protons and helium nuclei in a nuclei ratio in the range of about 12:1.

The paper explains that the Big-Bang dispersed relativistic protons and helium nuclei became the mysterious dark matter that now represents about 83 percent of the mass of the universe. Thus, the dark matter of the universe would also have a very low entropy and would also satisfy the Second Law of Thermodynamics.

The paper also explains that a relativistic-proton Big Bang would be a very efficient way of creating a universe and conserving its energy because the fewest number of particles having the most amount of usable energy would be created and dispersed.

On the other hand, a Big Bang creating Cold Dark Matter WIMPs, representing 83 percent of the mass of the universe, would be producing matter having a low percent of useful energy and high entropy, which would represent a very inefficient Big Bang design concept and a low-energy future for the universe.

The above-mentioned scientific paper, posted on the Cornell University Library arXiv.org physics website, is entitled "A Relativistic-Proton Dark Matter Would Be Evidence the Big Bang Probably Satisfied the Second Law of

Thermodynamics"[13]. It is dated February 15, 2007 and is available at: http://arxiv.org/abs/physics/0702132.

(This scientific paper[13] has played key roles in confirming the existence of both the Relativistic-Proton Dark Matter and the *Relativistic Big Bang*, and in providing an insight into *cosmic inflation*, and in the development of *Postmodern Cosmology*.)

CHAPTER 16

COSMIC NEWSWIRE #15

Dark Matter/Big Bang Mysteries May Have Been Solved Last Week at a London Cosmology Conference

LOS ALTOS HILLS, Calif., April 2, 2007 (AScribe Newswire) -- There are many cosmological mysteries of and questions about the physical universe that are studied by astronomers, astrophysicists, and cosmologists. These scientists have established the so-called "Standard Cosmological Model" that is the most complete description of the physical universe, which they utilize, study, and also question from time to time.

The scientists held a conference ended March 29, 2007 at the Imperial College, London, entitled "Outstanding Questions for the Standard Cosmological Model", sponsored by the U.S. National Science Foundation and the University of Alabama at Huntsville.

Jerome Drexler submitted a scientific paper to the cosmological conference. The paper appears to provide solutions to two important mysteries, namely, "What is the precise nature of the dark matter of the universe?" and "Did the Big Bang satisfy the Second Law of Thermodynamics?"

Drexler provided solutions to both of these mysteries in his cosmological paper entitled "A Relativistic-Proton Dark Matter Would Be Evidence that the Big Bang Probably Satisfied the Second Law of Thermodynamics"[13]. The paper is presented in Chapter 12.

(This research paper[13] turned out to be very useful in obtaining a physical insight into *cosmic inflation*, which is described as exhibiting a hyper-rapid, extremely short, exponential-growth period of the universe that took place a fraction of a second after the Big Bang. According to scientists, "the detailed particle physics mechanism responsible for cosmic inflation is not known". Drexler may have overcome that psychological barrier by deriving a solution to an important cosmic-inflation enigma, as described in Chapter 31.)

CHAPTER 17

Local Group Galaxy Cluster Dark Matter May Emit Extreme Ultraviolet and Infrared Photons

Extreme ultraviolet (EUV) from space is difficult to detect using an Earth-based telescope because the atmosphere scatters/absorbs a very large percentage of such light. A satellite-based EUV telescope could work.

Thus, although astronomers and cosmologists have assumed for the past 23 years that the dark matter of the universe is cold, passive, and absolutely dark it actually may be hot, active, and emit EUV or UV photons or soft X-rays.

Dark matter's relativistic protons orbiting a group of spiral galaxies in a cluster are a much more likely source of EUV or soft X-ray synchrotron emission than dark matter's relativistic protons in the dark matter halo of a single spiral galaxy, like the Milky Way. There are three reasons for this:

(1) A relativistic proton's synchrotron emission power is proportional to the square of its energy, (2) the wavelength of the peak emission power is inversely proportional to the square of the protons' energy, and (3) the energies of the dark matter protons in the Local Group galaxy cluster are

estimated at 30 times greater than those in the Milky Way's dark matter halo. (See Chapter 2006-50 located in Appendix C.)

Thus, dark matter relativistic protons orbiting groups of galaxies in the Local Group galaxy cluster should radiate synchrotron emission power about 900 times higher, at a wavelength 900 times smaller, than from protons in the Milky Way's dark matter halo.

Calculations indicate that the synchrotron emission photons from the Milky Way's dark matter halo should have a broad peak in the infrared including the wavelength of 5 microns. Photons from protons orbiting groups of galaxies in the Local Group galaxy cluster should have an EUV/soft X-ray broad peak including the wavelength of 5.5 nanometers.

The EUV-from-dark-matter theory could be tested in 2009 after the Hubble Space Telescope's EUV/UV sensitivity is increased by a factor of 10 in Aug 2008[20, 46], (see Chapter 33). The detection of UV, EUV or soft X-rays from the Local Group's dark matter would confirm Drexler's relativistic-proton dark matter hypothesis.

CHAPTER 18

COSMIC NEWSWIRE #16

Dark Matter's High Erosion Rate in Galaxy Clusters May Cause Universe's Accelerating Expansion

LOS ALTOS HILLS, Calif., May 7, 2007 (AScribe Newswire) -- In an expanding universe galaxy clusters are separating from one another according to Hubble's law. When the galaxy-cluster separation velocities continually speed up, as in our universe, the universe is in an accelerating expansion mode.

Since 1998, this has been attributed to a mysterious *dark energy*, called "the most profound mystery in all of science" by University of Chicago cosmologist Michael Turner.

In 1929, Edwin P. Hubble announced from the Mount Wilson Observatory near Los Angeles that with the exception of the galaxies closest to the Milky Way, galaxies are rushing away from each other in all directions and, therefore, the universe is expanding. His astronomical evidence led to the Big Bang theory of the creation of the universe.

In 1998, Saul Perlmutter, of the Lawrence Berkeley National Laboratory, completed a ten-year astronomical measurements study of events involving exploding stars (supernovae). This led to the discovery that the expansion of the universe was accelerating. These measurements were made by ground-based telescopes. Recently, Type 1a supernovae studies with the Hubble Space Telescope (HST) confirmed and refined the 1998 accelerating-expansion conclusions.

The cause was attributed to the so-called *dark energy*, a hypothetical form of energy that may permeate all space and may have a negative pressure resulting in a repulsive gravitational force. In 2001, Michael Turner essentially removed the word *all* from this definition when he wrote, "Dark energy by its very nature is diffuse and a low-energy phenomenon. It probably cannot be produced at accelerators; it isn't found in galaxies or even clusters of galaxies."[37]

Astro-cosmology author Jerome Drexler also believes that there is no evidence of the separation velocities accelerating between galaxies or between stars. That is, Drexler's criteria for the mysterious repulsive dark energy are: It accelerates the separation velocities between galaxy clusters, but has little or no effect on the separation velocities between galaxies or between stars.

His theory explains that in a first phase, galaxy clusters filled with dark matter protons are separating from each other with separation velocities proportional to their separations, according to Hubble's law.

Then, when galaxy clusters' relativistic dark matter protons *erode relativistic mass* at a high enough rate, via synchrotron emission of extreme ultraviolet (EUV), UV or soft X-ray photons, the separation velocities of galaxy clusters will increase because of the reduction in their mass lowering the gravitational attraction between them and owing to the Law of Conservation of Linear Momentum. (Synchrotron emission is explained below.)

This increases cluster-to-cluster separation and further lowers the gravitational attraction between galaxy clusters, thereby further accelerating both cluster separation and the expansion of the universe. The force causing this accelerating separation of galaxy clusters is essentially only between galaxy clusters, not between galaxies, as will be explained.

After a five-year analysis and interpretation of astronomical data, Drexler has concluded that dark matter is comprised of relativistic protons accompanied by relativistic helium nuclei in a nuclei ratio in the range of about 10:1 to 12:1. Note that it is widely believed that over 99 percent of the ordinary matter of the universe is comprised of hydrogen and helium in the atomic ratio of 12:1.

Galactic and extragalactic magnetic fields cause the charged dark matter particles to remain within galaxy clusters and to emit radiant energy in the form of extreme ultraviolet (EUV), soft X-ray, or infrared photons.

This is called synchrotron emission, which is photon emission from a proton in its direction of motion. Since protons in a galaxy cluster are orbiting groups of galaxies, such emission from a galaxy cluster should be relatively isotropic with respect to the galaxy cluster's linear motion. Therefore to a first approximation, synchrotron emission should not directly affect the cluster's linear motion significantly. However, dark matter protons, orbiting groups of galaxies, in galaxy clusters would be emitting relatively high power synchrotron emission, causing these protons to lose relativistic mass continuously at a high rate, as if their mass were eroding. Previous paragraphs in this chapter explain how this relativistic mass erosion leads to an acceleration of separation velocities between galaxy clusters.

How can synchrotron emission push galaxy clusters apart without pushing galaxies apart? First of all, the power of synchrotron emission from a relativistic proton in a magnetic field is directly proportional to the square of the proton's energy. Secondly, the energies estimated for the dark matter protons orbiting groups of galaxies within a typical galaxy cluster would be about 30 times higher than the energies calculated for the dark matter protons orbiting a single typical spiral galaxy, such as the Milky Way.

Thus, the synchrotron emission power per proton and mass loss rate per proton are about 900 times greater for pushing galaxy clusters apart than for pushing galaxies apart.

Another question is why the accelerating expansion did not begin until about five billion years ago, as reported by astronomers. Perhaps prior to that time the much smaller separations of galaxy clusters and the inverse-square law of gravity led to a very much higher gravitational attraction that minimized the synchrotron-radiation repulsive effect. Note that in the past 13.4 billion years galaxy-cluster separation distances have grown by a factor of the order of 1000.

Drexler's theories for dark matter and the accelerating expansion of the universe could be tested by NASA in 2009 when the Hubble Space Telescope's (HST's) UV (and EUV) sensitivity is increased by a factor of 30 (now estimated at 10). The detection of EUV or UV photons or soft X-rays from the dark matter of our Local Group galaxy cluster would confirm Drexler's relativistic-proton dark matter theory.

This is because calculations indicate that synchrotron emission from the Milky Way's halo should have a broad peak in the infrared while synchrotron emission from protons orbiting several to tens of galaxies in galaxy clusters should have a broad peak in the UV, EUV or soft x-ray region. (The NASA upgrade of the HST is now scheduled for August 2008[20, 46] instead of spring 2008. Also, see Chapter 33.)

CHAPTER 19

COSMIC NEWSWIRE #17

'Ring of Dark Matter' Uncovered from Anomalies/ Discrepancies, Says Cosmos Author Jerome Drexler

LOS ALTOS HILLS, Calif., May 14, 2007 (AScribe Newswire) -- The 'Ring of Dark Matter' was uncovered from the NASA-Hubble 3D Dark Matter Map data after astro-cosmology author Jerome Drexler pointed out on January 16, 2007 that the anomalies and discrepancies reported by astronomers on January 7 were actually valid data. NASA issued the following news advisory May 10.

"GREENBELT, Md. - NASA will hold a media teleconference at 1 p.m. EDT on May 15 to discuss the strongest evidence to date that dark matter exists. This evidence was found in a ghostly ring of dark matter in the cluster CL0024+17, discovered using NASA's Hubble Space Telescope. The ring is the first detection of dark matter with a unique structure different from the distribution of both the galaxies and the hot gas in the cluster. The discovery will be featured in the June 20 issue of the Astrophysical Journal." (A paper was also published May 2007 online as arXiv:

0705.2171[24], titled "Discovery of a Ringlike Dark Matter Structure in the Core of the Galaxy Cluster Cl0024+17".)

In early January 2007 the original research paper, "Dark matter maps reveal cosmic scaffolding," was presented and a press conference was held. The researchers stated their concerns about anomalies and discrepancies they had observed, "*naked* clumps of dark matter where there were no galaxies." Ten such statements are listed in a BBC NEWS article "Hubble makes 3D dark matter map"[36].

The researchers could not comprehend the existence of dark matter alone without galaxies. Jerome Drexler sensed that they must be relying on the unproven bottom-up theory of galaxy formation. He felt they should utilize his top-down theory of galaxy formation, in which dark matter can exist alone. With that in mind, he distributed the following Jan. 16, 2007 AScribe Newswire to the media and to 1000 dark-matter and NASA astronomers, astrophysicists, and cosmologists. Apparently it worked.

So-Called Anomalies in NASA-Hubble
3D Dark Matter Map are Explained by
Astro-Cosmology Author Jerome Drexler

LOS ALTOS HILLS, Calif., Jan. 16, 2007 (AScribe Newswire) -- The recent 3D mapping of the dark matter of the universe is a major astronomical accomplishment of

NASA and the Hubble telescope. However, the researchers' reports of so-called discrepancies and anomalies in the distribution of dark matter relative to the distribution of ordinary matter has prevented the 3D mapping from being an immediate cosmological success.

A study of the researchers' Jan. 7 comments during a press conference following the presentation of their scientific paper at the 209th meeting of the American Astronomical Society (AAS) in Seattle, Wash., provides some insight into the reasons behind the researchers' reports of apparent discrepancies and anomalies. (See their comments below.)

Based upon their stated concerns about the so-called discrepancies and anomalies, the researchers apparently believe in the bottom-up theory of galaxy formation, not the top-down theory of galaxy formation, which has some support including that of Jerome Drexler.

The astronomical data the researchers are concerned about actually support the top-down theory of galaxy formation and if the data had been described in that manner there would not have been the issue of discrepancies or anomalies — and the 3D mapping might have been considered a cosmological success as well as an astronomical success.

Some relevant comments by some of the dark-matter-map researchers at an AAS press conference in Seattle on Jan. 7, 2007 are as follows:

"We have to resolve discrepancies in the otherwise strong connection between ordinary matter and dark matter."

"Finding what I would call 'naked' clumps of dark matter where there are no galaxies for me is very strange. All dark matter clumps of sufficient size should have galaxies — if our understanding is correct."

"We see that dark matter concentrations sometimes seem to have no corresponding ordinary matter."

"For the moment, no one is talking about needing to revise cosmological models; but everything hinged on the size of these anomalies."

"The discrepancies could turn out simply to be artifacts, caused by noise in the data. But then again, they could be real."

A researcher said the anomalies were *tantalizing* and that his team was eager to investigate them more closely.

The bottom-up theory, where small galaxies form first and then merge over time to form large galaxies, is known to have serious flaws. For example, note that at the same AAS conference, New York University researchers gave a paper based upon their astrophysics paper, "A New Force in the Dark Sector?"[25] Their paper states, "The number of superclusters observed in SDSS data appears to be an order of magnitude [about ten times] larger than predicted by Lambda-Cold Dark Matter simulations."

Since Lambda-Cold Dark Matter simulations are based on the bottom-up theory of galaxy formation, the NYU researchers are indicating that the bottom-up theory is extremely inaccurate in predicting the number of superclusters in the universe. Therefore the so-called anomalies and discrepancies reported by the 3D dark-matter-map researchers may have evolved simply because the researchers tested the 3D astronomical data against the wrong galaxy-formation theory. The NASA-Hubble researchers should now test the so-called discrepancies and anomalies against the top-down theory of galaxy formation.

In the top-down theory of galaxy formation the well-known long, large, slightly curved dark matter filaments form potential galaxy clusters where such dark matter filaments intersect and collide[18]. Then small and large galaxies form in this potential-cluster region from the *remnants* of the collisions of the intersecting dark matter filaments. (In the case of Relativistic-Proton Dark Matter, the *remnants* are the useful protons and helium nuclei.)

The top-down theory of galaxy formation is further explained in the pages of ten Index references in Drexler's

May 2006, 295-page astro-cosmology book[8] titled *Comprehending and Decoding the Cosmos: Discovering Solutions to Over a Dozen Cosmic Mysteries by Utilizing Dark Matter Relationism, Cosmology, and Astrophysics.*

(This ends the January 16, 2007 newswire, which was sent to the media and to 1000 dark-matter and NASA astronomers, astrophysicists, and cosmologists.)

CHAPTER 20

COSMIC NEWSWIRE #18

Science Magazine's Dark Matter, 'Hydrogen in Some Hard-to-Track Form', Opens Door to Relativistic-Proton Dark Matter

LOS ALTOS HILLS, Calif., May 29, 2007 (AScribe Newswire) -- The scientific paper, "Missing Mass in Collisional Debris from Galaxies" *[27]* in the May 25 issue of Science magazine is significant in that it questions the 23-year-old mainstream Cold Dark Matter (CDM) theory, and it also opens the door of scientific acceptance to the competing five-year-old Relativistic-Proton Dark Matter cosmology.

The researchers' conclusion, a significant departure from Cold Dark Matter theory, reads: "it more likely indicates that a substantial amount of dark matter resides within the disks of spiral galaxies. The most natural candidate is molecular hydrogen in some hard-to-trace form."

The researchers point out that their conclusions disagree with the Cold Dark Matter theory that posits that (1) there is no dark matter in the disks of spiral galaxies and (2) that dark matter is comprised of non-baryonic matter, which excludes hydrogen and protons.

In agreement with the researchers' conclusion is the five-year-old competing Relativistic-Proton Dark Matter theory and cosmology that posits that relativistic-protons, a hard-to-trace form of hydrogen, does indeed reside within the disks of spiral galaxies, as well as in their halos.

The Science paper clearly establishes new constraints on the nature and location of dark matter in spiral galaxies and in recycled-from-debris dwarf galaxies. The paper carefully analyzes astronomical dark matter in a triplet of recycled dwarf galaxies formed from debris from the collision of two massive spiral galaxies. The current mainstream Cold Dark Matter theory hypothesizes that such recycled-from-debris dwarf galaxies should be free of non-baryonic dark matter.

However, all three of the recycled dwarf galaxies were discovered to have twice as much dark matter as ordinary matter. Therefore the researchers were forced to conclude that the dark matter in debris-based dwarf galaxies must be baryonic since it could not be non-baryonic. They further concluded that the recycled dwarf galaxy's baryonic dark matter would have come from the disks of the colliding massive spiral galaxies.

The researchers' conclusion that the disks of spiral galaxies harbor "molecular hydrogen in some hard-to-trace-form" opens the door of scientific acceptance to the five-year-old Relativistic-Proton Dark Matter cosmology. This relatively new dark-matter cosmology is described in two recently published books[5,8] and in two recent scientific papers[12,13], all authored by Silicon Valley's J.Drexler.

The abstract of the Science paper reads as follows: "Recycled dwarf galaxies can form in the collisional debris of massive galaxies. Theoretical models predict that, contrary to classical galaxies, these recycled galaxies should be free of non-baryonic dark matter. By analyzing the observed gas kinematics of such recycled galaxies with the help of a numerical model, we demonstrate that they do contain a massive dark component amounting to about twice the visible matter. Staying within the standard cosmological framework, this result most likely indicates the presence of large amounts of unseen, presumably cold, molecular [hydrogen] gas. This additional mass should be present in the disks of their progenitor spiral galaxies, accounting for a substantial part of the so-called missing baryons." (Science 25 May 2007 Vol.316 no.5828, pp.1166-1169)

Jerome Drexler originated the five-year-old Relativistic-Proton Dark Matter theory and cosmology and disclosed and defended it in the form of a 32-slide Powerpoint presentation to two professors at the University of California, Santa Cruz campus in April 2003. He then expanded his presentation to 108 slides and transformed it into a 156-page paperback book, *How Dark Matter Created Dark Energy and the Sun,* which was published December 15, 2003.

Drexler followed this with a 19-page scientific paper in April 2005, published on the physics website arXiv.org as e-Print No. astro-ph/0504512[12], a five-page scientific paper in February 2007, as e-Print No. physics/0702132[13], and a 295-page paperback book entitled *Comprehending and Decoding the Cosmos* published May 22, 2006.

CHAPTER 21

COSMIC NEWSWIRE #19

Eroding High-Energy Dark Matter Particles in Galaxy Clusters May Explain the Universe's Acceleration

LOS ALTOS HILLS, Calif., July 19, 2007 (AScribe Newswire) -- The 2006 Shaw Prize in Astronomy and 2007 Gruber Cosmology Prize were awarded for the amazing discovery of the accelerating expansion of the universe, but the cause of this phenomenon is still considered a mystery.

Dark matter comprises about 83 percent of the mass of the universe. If dark matter were comprised of uncharged particles, the author would have no explanation for the accelerating expansion of the universe.

Ordinary matter of the universe is comprised of over 99 percent in the form of hydrogen and helium in the well-known atomic ratio of 12:1. The author, over the past eight years, has accumulated overwhelming evidence that the dark matter of the universe is comprised of relativistic protons and helium nuclei in the nuclei ratio about 10:1 to 12:1 that orbit galaxies and groups of galaxies. The proton orbits are not determined by gravitational attraction as in the case of our solar system, but by the extragalactic magnetic field and the electric charges and the velocities of the protons and helium nuclei. More exactly, the proton orbits are determined by the Larmor radius equation rather than Kepler's laws, which apply to the solar system. (See Appendix B, pages 213-215)

Based upon the overwhelming evidence provided by the author's astro-cosmology books of 2003, 2006, and 2008 and his scientific papers of 2005 and 2007, the reader is asked to proceed, at the moment, on the basis that the author's dark matter theory is probably valid. The following is the only plausible explanation, publicly proposed to date, for the mystery of the accelerating expansion of the universe.

Observed galaxy clusters are rushing apart faster and faster thereby accelerating the expansion of the universe. If the galaxies and the stars were similarly rushing apart faster and faster, then the *hypothetical* concept of *dark energy* pervading all space might be valid. But this isn't the case.

The galaxy clusters are indeed rushing apart as their separation velocities are accelerating, but the separation velocities of the galaxies and of the stars are not known to be accelerating. If the hypothetical pervasive *dark energy* actually existed, probably all three of these celestial bodies would be rushing apart faster and faster.

In an expanding universe galaxy clusters are moving away from one another according to Hubble's law. When the galaxy-cluster separation velocities continue to rise, as in our universe, the universe is in an accelerating expansion mode. Since 1998, this has been attributed to *dark energy*, called "the most profound mystery in all of science" by University of Chicago cosmologist Michael Turner.

In that year, two astronomical research studies led by Saul Perlmutter, of the Lawrence Berkeley National Laboratory and by Brian Schmidt of the Australian National University, of astronomical events involving exploding stars (supernovae) led to the award-winning discovery that the expansion of the universe was accelerating. The cause was attributed to *dark energy*, a hypothetical form of energy assumed to permeate *all* space and to have negative pressure resulting in a repulsive gravitational force.

In 2001, Michael Turner essentially removed the word *all* from this definition when he wrote, "Dark energy by its very nature is diffuse and a low-energy phenomenon. It probably cannot be produced at accelerators; it isn't found in galaxies or even clusters of galaxies"[37]. Astro-cosmology author Jerome Drexler has based his accelerating universe theory on Michael Turner's statement, "it isn't found in galaxies or even clusters of galaxies." That is, Drexler's criteria for the mysterious repulsive force are: it pushes galaxy clusters apart faster and faster, but it doesn't push galaxies or stars apart faster and faster.

Drexler's theory explains that, in a first phase, galaxy clusters, filled with relativistic dark matter protons orbiting groups of galaxies, are separating from each other with velocities proportional to their separations according to Hubble's law. Then, when galaxy clusters' relativistic dark matter protons erode relativistic mass at a sufficiently high rate, via synchrotron emission of extreme ultraviolet (EUV) or soft X-ray photons, the separation velocities of galaxy clusters will increase because of the reduction in the gravitational attraction between them and owing to the Law of Conservation of Linear Momentum. (The "sufficiently high rate" should be determined by the gravitational attraction between galaxy clusters versus the level of synchrotron emission from the dark matter protons.)

This increases cluster-to-cluster separation and further lowers the gravitational attraction between galaxy clusters, thereby further accelerating both galaxy cluster separation and the expansion of the universe. The force causing this accelerating separation of galaxy clusters is essentially only between galaxy clusters, not between galaxies, or between stars, as will be explained.

Drexler's dark matter candidate is comprised of relativistic protons and relativistic helium nuclei in a ratio of between

10:1 and 12:1. This is compatible with a recent scientific paper "Missing Mass in Collisional Debris from Galaxies" in the May 25, 2007 issue of Science magazine[27], which concludes, "The most natural [dark matter] candidate is molecular hydrogen in some hard-to-trace form."

The extragalactic magnetic fields cause the charged ultra-high-energy dark matter protons to remain within galaxy clusters and to radiate synchrotron emission primarily in the form of extreme ultraviolet (EUV), UV, or soft X-ray, or infrared photons.

Synchrotron emission from a proton is in the form of photons in its direction of motion. Since protons in a galaxy cluster are orbiting groups of galaxies, such emission from a galaxy cluster should be relatively isotropic with respect to the cluster's linear motion. The higher energy relativistic protons, orbiting groups of galaxies, would be emitting higher power synchrotron emission, causing them to lose energy and relativistic mass faster, as if their mass were eroding faster.

How can synchrotron emission push galaxy clusters apart without pushing galaxies apart? First of all, the power of synchrotron emission from a relativistic proton moving across a transverse magnetic field is directly proportional to the square of the proton's energy. Secondly, the energies estimated for the dark matter protons orbiting groups of galaxies in a galaxy cluster would be about 30 times higher than the energies calculated for the dark matter halo protons orbiting a single spiral galaxy, such as the Milky Way. (See Chapter 2006-50, Appendix C.)

Thus, the synchrotron emission power per proton and relativistic mass loss rate per proton are about 900 times greater for accelerating the separation velocities of galaxy clusters than for accelerating the separation velocities of galaxies.

Another question is why the accelerating expansion did not begin until about five billion years ago, as reported by astronomers. Perhaps in earlier epochs the much smaller separations of galaxy clusters and the inverse-square law of gravity led to a very much higher gravitational attraction between clusters that minimized the synchrotron-emission repulsive effect. Note that in the past 13.4 billion years galaxy-cluster separation distances have grown by a factor of the order of 1000.

Drexler's theories for dark matter and the accelerating cosmos could be tested by NASA in 2009 (and Russia in 2010[14]) when the Hubble telescope's EUV/UV sensitivity is increased[20,46] by a factor of 10. The detection of EUV or UV photons or soft X-rays from dark matter of our Local Group galaxy cluster would support Drexler's Relativistic Proton Dark Matter theory.

Calculations indicate that synchrotron emission from the Milky Way's dark matter halo should have a broad peak in the infrared, while synchrotron emission from dark matter protons orbiting groups of galaxies in galaxy clusters should have a broad peak in the EUV or soft X-ray region.

Is the dark matter of the universe comprised of ultra-high-energy relativistic protons? Consider these five paragraphs.

The only plausible explanation for the *accelerating expansion of the universe*, announced to date, requires that the dark matter of the universe be comprised of ultra-high-energy relativistic protons (see Chapter 18).

The only plausible explanation for the *Big Bang satisfying the Second Law of Thermodynamics*, announced to date, requires that dark matter be comprised of ultra-high-energy relativistic protons (see Chapter 12).

The only announced plausible explanation for *UHE cosmic ray protons* with energies above 60 EeV bombarding Earth's atmosphere requires that dark matter be comprised of ultra-high-energy relativistic protons (see Chapters 29 and 31).

The only announced plausible explanation for *cosmic inflation* requires that dark matter be comprised of ultra-high-energy relativistic protons (see Chapters 31 and 34).

The only announced plausible explanation for the *cosmic web* requires that dark matter be comprised of ultra-high-energy relativistic protons (see Chapter 33).

CHAPTER 22

COSMIC NEWSWIRE #20

Relativistic-Proton Dark Matter Boosted Over Cold Dark Matter in 2007 via Three Scientific Papers and a Scientific Essay

LOS ALTOS HILLS, Calif., Aug. 8, 2007 (AScribe Newswire) -- A scientific paper authored by Professor Stacy McGaugh, "Seeing Through Dark Matter"[28], in the August 3 issue of Science magazine ends with a sentence welcomed by those who question the existence of Cold Dark Matter, "The universe may not be as cold and dark as we imagine." A sentence lending support to relativistic-proton dark matter over Cold Dark Matter reads, "The collision velocities of the components of the Bullet cluster are extraordinarily high ..."

McGaugh also comments, "the observations of Bournaud et al. may pose an existential crisis for non-baryonic (non-proton) dark matter." This refers a scientific paper, "Missing Mass in Collisional Debris from Galaxies"[27], published in the May 25 issue of Science magazine, which provides significant scientific support for Relativistic-Proton Dark Matter.

The third relevant scientific paper was published online February 15, 2007 on the physics arXiv designated physics/0702132 *[13]* that explains that the Big Bang would have satisfied the Second Law of Thermodynamics by creating Relativistic-Proton Dark Matter. The paper was authored by Drexler and provides the most significant 2007 scientific support for Relativisic-Proton Dark Matter over Cold Dark Matter and also opened the path to postmodern Big Bang cosmology.

Also in 2007, the reputation of the Relativistic-Proton Dark Matter theory was enhanced by a plausible explanation for the accelerating expansion of the universe On July 19, Drexler wrote an essay entitled, "Eroding High-Energy Dark Matter Particles in Galaxy Clusters May Explain the Universe's Acceleration", that became Chapter 21.

Returning to Science magazine's May 25 paper, the researchers' conclusion reads: "it more likely indicates that a substantial amount of dark matter resides within the disks of spiral galaxies. The most natural candidate is molecular hydrogen in some hard-to-trace form."

The researchers point out that their conclusions disagree with the Cold Dark Matter theory that posits that there is no dark matter in the disks of spiral galaxies and also that dark matter is comprised of non-baryonic matter, which excludes hydrogen and protons.

However, in agreement with the researchers' conclusion is Drexler's competing Relativistic-Proton Dark Matter theory and cosmology that posits that relativistic protons, a hard-to-trace form of hydrogen, does reside within the disks of spiral galaxies, as well as in their halos.

The May 25 Science paper clearly establishes new constraints on the nature and location of dark matter in spiral galaxies and in recycled-from-debris dwarf galaxies. The paper carefully analyzes astronomical dark matter in a triplet of recycled dwarf galaxies formed from debris from the collision of two massive spiral galaxies. The Cold Dark Matter theory indicates that such recycled-from-debris dwarf galaxies should be free of non-baryonic (non-proton) dark matter.

It turns out that all three of the recycled dwarf galaxies were discovered to have twice as much dark matter as ordinary matter. The researchers were forced to conclude that the dark matter in debris-based dwarf galaxies must be baryonic (proton-based) since it could not be non-baryonic. They further concluded that the recycled dwarf galaxy's baryonic (proton) dark matter probably came from the *disks* of the colliding massive spiral galaxies.

Thus, the researchers' conclusion that the *disks* of spiral galaxies harbor "molecular hydrogen in some hard-to-trace-form" provide double-barreled support for Relativistic-Proton Dark Matter over Cold Dark Matter.

CHAPTER 23

COSMIC NEWSWIRE #21

Explaining NASA Dark Matter Enigma: Abell 520 versus the Bullet Cluster via *Comprehending and Decoding the Cosmos*

LOS ALTOS HILLS, Calif., Aug. 23, 2007 (AScribe Newswire) -- NASA issued a news release on August 16, 2007 entitled, "Dark Matter Mystery Deepens in Cosmic Train Wreck"[29]. It describes the unusual features of the merger-formed Abell 520 complex galaxy cluster comprising a small group of bright galaxies devoid of dark matter, a massive region of dark matter, and a region of X-ray emitting gas. Abell 502 represents the aftermath of a collision-merger between two equal-size galaxy clusters.

The collision-merger of Abell 520 apparently caused dark matter to separate from the two merging galaxy clusters and to combine to form a massive region of dark matter with a scattering of galaxies surrounding it and an isolated small region containing a small group of bright galaxies that appears to be devoid of dark matter.

The August 2007 galaxy-cluster-collision mystery of Abell 520 is very different from the August 2006 report of the collision-formed Bullet Cluster (also known as 1E 0657-56).The Bullet Cluster was created by a small low-mass galaxy cluster passing through a larger galaxy cluster at about 4500 km/sec, yielding a merger time of about one billion years. Some researchers consider the Bullet Cluster's collision velocity to be "very high" and not ordinary.

The mass of the smaller cluster, that merged to form the Bullet Cluster, was about 22 percent of the mass of the larger one. Its mass was so low and its velocity was so high that the effective interaction of the dark matter/galaxies of the small cluster on the larger cluster apparently was not substantial. This is evident from the fact that in the case of the Bullet Cluster each of the original galaxy clusters has retained its own dark matter and its own bright galaxies, even after the merger of clusters was completed.

On the other hand, Abell 520 was created by the merger of two equal-size galaxy clusters at a collision velocity probably lower than the "very high" 4500 km/sec. The collision velocity of Abell 520 apparently was low enough and its two cluster masses high enough to permit substantial interaction between the dark matter/galaxies of the merging galaxy clusters. The appearance of the post-merger massive dark matter region in Abell 520 attests to a true galaxy cluster merger having taken place, compared to the brief-encounter-merger appearance of the Bullet Cluster, which still carries the images of the two original galaxy clusters.

The composite multi-color images shown in Fig.2 of the scientific paper "A Dark Core In Abell 520", arXiv: 0706.3048v1[39], exhibits the extreme interaction characteristics that might be expected of a collision-merger of galaxy clusters involving the creation of multiple starburst galaxies, which are active and dynamic and have high star formation rates (SFRs) 50 times higher than normal.

Under the Relativistic-Proton Dark Matter theory a spiral galaxy's dark matter halo is consumed over time in the star formation process and in the hydrogen fusion process within stars (This is true for a spiral galaxy or a starburst galaxy.) while the higher-energy relativistic dark-matter protons orbiting groups of galaxies would tend to be ejected during galaxy cluster mergers involving starburst galaxies. Thus, after Abell 520's one billion-year cluster-merger process there could be regions of only dark matter, of bright spiral galaxies devoid of dark matter halos, of bright spiral galaxies with diminished halos, and possibly some burned-out dark galaxies — a composite cosmic "train wreck."

A plausible explanation for the make-up of Abell 520 may be derived via a study of Chapters 2006-50, 2006-47 and 2006-46 of *Comprehending and Decoding the Cosmos*[8]. They are available in Appendix C. This same information formed the theoretical basis for the interaction phenomena described on the previous page. Comments about these chapters follow.

Chapter 2006-50[8], entitled "Different Dark Matter for Small Galaxies and for Clusters", discusses Relativistic-Proton Dark Matter having two different kinetic energy peaks (designated dual dark matter) within a galaxy cluster. It also discusses self-interaction by dark matter.

Chapter 2006-47[8], entitled "UHECR Protons via Starburst Galaxies/ Merging Galaxies", focuses on the ejection of the higher-energy dark matter protons from starburst galaxies, which are created by the merging of two galaxies from two merging galaxy clusters. The chapter references two 2005 scientific papers by other researchers as support for the theory. (UHECR stands for Ultra-high-energy cosmic rays.)

Chapter 2006-46[8], entitled "Starburst Galaxies Form via Merging Galaxy Clusters", explains how some galaxy cluster mergers trigger the formation of highly active and dynamic starburst galaxies, which theoretically could create an Abell 520-like cosmic "train wreck" over a long merger time scale.

The Relativistic-Proton Dark Matter theory posits that a spiral galaxy's dark matter halo is comprised of galaxy orbiting relativistic protons whose orbits are determined by the proton kinetic energy and the strength of the transverse extragalactic magnetic field. Within a galaxy cluster, in addition to dark-matter protons orbiting individual spiral galaxies, there are also higher-energy relativistic dark-matter protons in larger orbits that are orbiting groups of several to tens of galaxies.

Chapter 2006-50 points out that in the Local Group galaxy cluster the average kinetic energy of these higher-energy protons is estimated to be about 30 times higher than that of the protons orbiting a single spiral galaxy. The quantity of these higher-energy protons (whose synchrotron emission is in the extreme UV or soft X-ray range) in a galaxy cluster is estimated to be much smaller than those protons orbiting single spiral galaxies (whose synchrotron emission is probably in the infrared).

The orbits of both the higher-energy protons and the lower-energy protons are determined by the transverse extragalactic magnetic field they encounter and their energies. During the merger of two galaxy clusters, magnetic shocks are produced when the magnetic fields are distorted or altered. The level of magnetic shock depends upon the nature of each of the two galaxy clusters, their relative orientation, the degree of merger completion, their collision velocity, the time period of interaction, and other features of the cluster merging process.

Such magnetic field shocks should cause the higher-energy higher-velocity dark matter protons to be ejected from their orbits as reported in Chapter 2006-47 of the May 2006 book[8]. If the magnetic field shocks were more severe, then some of the lower-energy lower velocity protons orbiting single galaxies also could be ejected from their orbits.

On the basis of this dual-dark-matter explanation, perhaps the Relativistic-Proton Dark Matter of the Bullet Cluster experienced a much smaller magnetic field shock during the lower-interaction merger of galaxy clusters than in the case of Abell 520. This might have permitted most or almost all of the dark matter protons to remain with the merging galaxies in the Bullet Cluster.

The more severe Abell 520 magnetic shocks might have caused essentially all of the higher-energy dark matter protons and some to all of the lower-energy dark matter protons to depart many of the bright galaxies of Abell 520.

This could explain the Abell 520 dark matter core being devoid of bright galaxies while having bright galaxies in another region being devoid of dark matter. Note that Chapter 2006-47 uses this same theory to explain that during mergers of galaxy clusters and their galaxies some of the higher-energy relativistic protons will be ejected and either will be added to the dark matter of the merged clusters or will enter some star systems as ultra-high-energy cosmic ray protons Two scientific papers reporting such observations are Diego F. Torres and Luis A. Anchordoqui, 2005, astro-ph/0505283 v1[40] and Elena Pierpaoli and Glennys Farrar, 2005, astro-ph/0507679 v3[41].

Perhaps it should be noted here that the reported Chandra X-ray photons detected from the very hot gas of Abell 520 (located near the dark matter core) actually could include some soft X-ray synchrotron emission photons from the ultra-high-energy relativistic dark matter protons.

CHAPTER 24

COSMIC NEWSWIRE #22

NASA Data Raises Doubts of Existence of Cold Dark Matter in Galaxy Clusters

LOS ALTOS HILLS, Calif., Sept. 10, 2007 (AScribe Newswire) -- A NASA Web site entitled *Dark Matter*[30] provides information about dark matter and galaxy clusters that raises doubts about the existence of Cold Dark Matter in galaxy clusters.

The NASA Web site says, "The space between galaxies in clusters is filled with a hot gas. In fact, the gas is so hot (tens of millions of degrees!) that it shines in X-Rays instead of visible light." And, "Remarkably, it turns out there is five times more material in clusters of galaxies than we would expect from the galaxies and hot gas we can see."

These two statements seem to imply that there are some unanswered questions regarding the hot gas, X-ray emission, dark matter, and galaxy clusters. One such question might be whether these two statements taken together rule out the existence of Cold Dark Matter in galaxy clusters. An argument providing such evidence might proceed as follows:

If Cold Dark Matter were, as indicated by NASA, four times as plentiful as the mass of both the galaxies and the hot gas in a galaxy cluster and had to share the space between galaxies with the super-hot gas, how could the gas reach a temperature of "tens of millions of degrees"? Wouldn't a massive volume of Cold Dark Matter, occupying the same space and in intimate contact with the gas, cool the gas and prevent it from reaching such an extremely high temperature? Also, how could the Cold Dark Matter remain cold under such circumstances?

Related questions are whether this NASA Web site information is known to dark matter researchers and whether those who are aware of it are properly considering it.

CHAPTER 25

COSMIC NEWSWIRE #23

Science Magazine's Warm Dark Matter Discovery Supports Warm-Hot Relativistic-Proton Dark Matter and Visa Versa

LOS ALTOS HILLS, Calif., Sept. 17, 2007 (AScribe Newswire) -- A paper in the September 14, 2007 issue of Science magazine "Lighting the Universe with Filaments"[31] claims that computer simulations disclose that Warm Dark Matter (WDM) would create enormous dark matter filaments that in turn would create sun-like long-life stars that could exist until now.

These simulations also show that Cold Dark Matter (CDM) would only create short-life high-mass stars, typically with a few hundred times the mass of the Sun that would have exploded billions of years ago.

Therefore, a future astronomical discovery of ancient first-generation sun-like stars would support the Warm Dark Matter theory over the Cold Dark Matter theory. This type of research represents a new approach to uncovering the nature of the dark matter of the universe.

Jerome Drexler, author of *Comprehending and Decoding the Cosmos*[8] devoted Chapter 2006-33 of this earlier book to the discovery and significance of the long, large dark matter filaments announced by NASA and Harvard on September 9, 2004[18]. The simulations described in the recent Science paper[31] seem compatible with the 2004 NASA/Harvard astronomical research data[18], which make the simulations more credible and more relevant.

Thus, the 2006 book's Chapter 2006-33 is presented here as a response. It is also a way of recognizing the important computer-simulation contributions of the authors of the Science magazine paper, Liang Gao and Tom Theuns:

Comprehending and Decoding the Cosmos, Chapter 33

Intersecting Dark Matter Filaments Create Galaxy Clusters

"Relativistic-Proton Dark Matter Particles Could Be Concentrated in the Long, Large Filaments of Dark Matter That Form Galaxy Clusters Where the Dark Matter Filaments Intersect"

"The September 9, 2004 news release from NASA (and Harvard) entitled, "Motions in nearby galaxy cluster reveal presence of hidden superstructure,"[18] regarding Chandra x-ray images of the Fornax cluster, states:"

"Astronomers think that most of the matter in the universe is concentrated in long large filaments of dark matter and that galaxy clusters are formed where these filaments intersect."

"The researchers' related paper (astro-ph/0406216) is entitled, "The Chandra Fornax Survey – I: The Cluster Environment"[42]. This astronomically established filamentary description of dark matter appears to be much more compatible with the Relativistic-Proton Dark Matter theory than the Cold Dark Matter theory. It seems highly unlikely that the dark matter (DM) filamentary structure could be created by very slow moving, weakly interacting (only through gravitational tidal forces) particles."

"The vision of DM filaments crisscrossing the cosmos gives the impression of high-velocity particles, while the crashing of intersecting DM filaments creating galaxy clusters gives the impression of a top-down theory of galaxy formation. Both of these impressions point toward and lend support to the Relativistic-Proton Dark Matter theory and cosmology."

"Furthermore, the theoretical WIMPs, being non-baryonic, cannot be transformed into hydrogen and helium where the DM filaments intersect, whereas the relativistic protons and helium nuclei, being baryonic, can provide hydrogen and helium to the galaxy clusters where the filaments intersect. For the above reasons, the September 2004 reports of the DM filaments seemed to be very supportive of Drexler's DM theory and encouraged him to write this *(2006)* book sequel" [8].

(Three and one-third years later these long large dark matter filaments have become associated with the *cosmic web*. Science magazine published a special issue on January 4, 2008 focusing on the universe's *cosmic web* the filamentary dark matter structure that appears to connect all galaxy groups and galaxy clusters. Adrian Cho's Science article "Untangling the Celestial Strings"[44] begins:

"Unraveling that *cosmic web* presents astronomers, astrophysicists, and cosmologists with their next great challenge — The [cosmic] web is the framework on which the universe is built. It consists primarily of "dark matter," mysterious stuff that makes up 85% of the matter in the universe but has revealed itself only through its gravity. Enormous filaments and blobs of the stuff condensed as the universe matured. Within them nestle the galaxies and their stars, creating streams of light stretching between inky voids."

Science article, "The Cosmic Web in Our Own Backyard"[47], begins with: "On the largest scales, matter is strung out on an intricate pattern known as the *cosmic web*. The tendrils of this web should reach right into our own cosmic backyard, lacing the Galactic halo with lumps of dark matter. The search for these lumps, lit up by stars that formed within them, is a major astronomical endeavor...")

CHAPTER 26

COSMIC NEWSWIRE #24

British Professor Elegantly Questions Validity of Cold Dark Matter Hypothesis

LOS ALTOS HILLS, Calif., Oct. 23, 2007 (AScribe Newswire) -- The September-October 2007 issue, Volume 95, of the American Scientist magazine, published a remarkable article by Michael J. Disney, an emeritus professor in the School of Physics and Astronomy at Cardiff University in the UK. The article[32] fully lives up to its title, "Modern Cosmology: Science or Folk Tale?"

Professor Disney uses Big Bang cosmology as the basis for his thesis and shows that the accepted mainstream Big Bang cosmology relies on too few astronomical observations and too many hypotheses to be considered "science". In the article he sometimes uses the term "free parameters" as a synonym for the word "hypotheses".

Professor Disney's negative opinion of the Cold Dark Matter (CDM) hypothesis in Big Bang cosmology is described in the following two paragraphs quoted from his article.

The Significance of Cosmology

"The currently fashionable concordance model of cosmology (also known to the cognoscenti as 'Lambda-Cold Dark Matter,' or 'Lambda-CDM') has 18 parameters, 17 of which are independent. Thirteen of these parameters are well fitted to the observational data; the other four remain floating. This situation is very far from healthy. Any theory with more free parameters [hypotheses] than relevant [astronomical] observations has little to recommend it. Cosmology has always had such a negative significance, in the sense that it has always had fewer [astronomical] observations than free parameters [hypotheses] (as are illustrated on page TK), though cosmologists are strangely reluctant to admit it. While it is true that we presently have no alternative to the Big Bang in sight, that is no reason to accept it. Thus it was that witchcraft took hold.

"The three successful predictions of the concordance model (the apparent flatness of space, the abundances of the light elements and the maximum ages of the oldest star clusters) are overwhelmed by at least half a dozen unpredicted surprises, including dark matter and dark energy. Worse still, there is no sign of a systematic improvement in the net significance of cosmological theories over time."

Silicon Valley's inventor/scientist, Jerome Drexler, raised the alarm about the validity of the Cold Dark Matter hypothesis in his December 15, 2003 published book *How*

Dark Matter Created Dark Energy and the Sun[5] which also offered a warm/hot form of dark matter as an alternative.

Only the British seemed to be interested in his 2003 book until Drexler authored the 295-page, "Comprehending and Decoding the Cosmos"[8], published May 22, 2006. During the past 15 months, more and more authors of cosmology related papers and articles have directly or indirectly questioned the validity of the Cold Dark Matter hypothesis. Has the 23-year-old Cold Dark Matter hypothesis evolved into the equivalent of a scientific mistake?

Recently, Science magazine published three papers or articles questioning the Cold Dark Matter hypothesis, namely, on May 25, "Missing Mass in Collisional Debris from Galaxies"[27], on August 3, "Seeing Through Dark Matter"[28], and on September 14, "Lighting the Universe with Filaments"[31].

Hopefully, Science magazine will be considered for a Pulitzer Prize in journalism for exposing the apparent mainstream Cold Dark Matter hypothesis that has evolved into a scientific mistake. Unfortunately this scientific mistake continues to retard progress in cosmology, demoralize cosmology researchers, and ill-prepares future cosmology researchers and yet, it continues to receive substantial financial support from U.S. government agencies and support from most U.S. universities.

CHAPTER 27

COSMIC NEWSWIRE #25

Thirteen Astronomical Observations
to Solve British Professor's
Dark Matter Conundrum

LOS ALTOS HILLS, Calif., Nov.1, 2007 (AScribe Newswire) -- The article "Modern Cosmology: Science or Folk Tale?"[32] in the September-October issue of American Scientist magazine and the Oct. 23 AScribe Newswire release "British Professor Elegantly Questions Validity of Cold Dark Matter Hypothesis" (now Chapter 26) point out that modern Big Bang cosmology based upon Cold Dark Matter relies upon too few astronomical observations and upon too many hypotheses to be considered *science*.

The problem with the 23-year-old Cold Dark Matter hypotheses is that if we attempt to add another astronomical observation to improve the net significance of the Big Bang cosmology results, we most probably would have to add another hypothesis to the list of Cold Dark Matter hypotheses for the group of them to be compatible with the new astronomical observation. Thus, there would be no improvement in net significance.

That is, the problem with modern Big Bang cosmology is not a shortage of available relevant astronomical observations, but that the Cold Dark Matter hypotheses cannot generally show compatibility with an added independent astronomical observation without adding another new hypothesis.

The three previous paragraphs are an attempt to describe the dark matter conundrum posed in "Modern Cosmology: Science or Folk Tale?"[32] by Professor Michael J. Disney of Cardiff University. This response endeavors to provide a scientific solution to Disney's dark matter conundrum, which appears to be fundamental to Big Bang cosmology.

Could two or three of the following 13 astronomical observations be used to improve the net significance of Big Bang cosmology theory by using the Cold Dark Matter hypotheses plus just one more hypothesis? Theoretically, that could improve the net significance of modern Big Bang cosmology if such a hypothesis could be found.

Thirteen Independent Astronomical Observations Related to Big Bang Cosmology

1. Hydrogen and helium in an atomic ratio of about 12:1 comprise over 99 percent of the mass of the universe.

2. Relativistic protons and relativistic helium nuclei, which are known to be common in the universe, are in a nuclei ratio estimated at about 10:1 near Earth.

3. Dark matter represents about 83 percent of the mass of the universe.

4. Astronomical data indicate the dark-matter particle mass is at least 35 times greater than the mass of a proton at rest.

5. Dark matter in early epochs was concentrated in long, large, curved filaments, announced by NASA[18] on September 8, 2004, which formed galaxy clusters where two dark matter filaments intersected.

6. The separation velocities between galaxy clusters began to accelerate about 5 billion years ago, whereas no acceleration of separation velocities between galaxies or between stars has been reported.

7. Strongly self-interacting dark matter particles are normally located in halos of dwarf and low surface brightness (LSB) galaxies while weaker self-interacting dark matter particles are normally located within galaxy clusters, orbiting groups of galaxies, implying the existence of two types, forms, or modes of dark matter particles.

8. Starburst galaxies created by the merging of spiral galaxy clusters exhibit extremely high star formation rates.

9. The blue and blue-white stars, as young as one million years old, are located in the spiral arms of mature spiral galaxies, which also contain five-billion-year-old red stars in their nuclei.

10. The Big Bang, as a thermodynamic process, satisfied the Second Law of Thermodynamics, which requires that the total amount of disorder, or entropy, in the universe,

always increases with time. Also, since the Big Bang represents the beginning of time for the universe its entropy should have been at its lowest level for all time.

11. The nuclei [cores] of starburst galaxies usually exhibit new blue star formation while spiral galaxies exhibit new blue star formation in their spiral arms.

12. Large and mature spiral galaxies were in existence less than 2.5 billion years after the Big Bang.

13. Authors of "Missing Mass in Collisional Debris from Galaxies"[27] conclude, "it more likely indicates that a substantial amount of dark matter resides within the disks of spiral galaxies. The most natural candidate is molecular hydrogen in some hard-to-trace form."

Consider these 13 astronomical observations. Could two or three of them be selected and utilized to improve the net significance of modern Big Bang cosmology theory by using the Cold Dark Matter hypotheses plus just one more hypothesis? If the answer is yes, it would make a very interesting scientific paper.

On the other hand, most of the 13 astronomical observations can be used as a group or separately to improve the net significance of postmodern Big Bang cosmology, without adding another hypothesis. This can be done by utilizing the Relativistic-Proton Dark Matter, which is described as comprising galaxy-orbiting relativistic protons accompanied by relativistic helium nuclei in a nuclei ratio in the range of about 10:1 to 12:1.

Why does Silicon Valley's inventor/scientist Jerome Drexler believe this is feasible?

Drexler has authored books published in December 2003 and in May 2006, scientific papers in April 2005 and February 2007 and about thirty AScribe scientific news releases, called cosmic newswires in this book, since June 2006. This collection provides strong evidence that dark matter is comprised of galaxy-orbiting relativistic protons accompanied by relativistic helium nuclei in a nuclei ratio in the range between 10:1 and 12:1.

Such relativistic protons and helium nuclei are common in the universe. When they bombard Earth's atmosphere they are called cosmic rays (and ultra-high-energy cosmic rays), which were discovered in 1912 by Victor F. Hess, who won the 1936 Nobel Prize in physics for the discovery.

Drexler first published his Relativistic-Proton Dark Matter theory in his December 2003 book[5], his advanced version of the theory, which entered into cosmology, in his May 2006 book[8], and his postmodern Big Bang cosmology theory in a February 2007 scientific paper[13].

CHAPTER 28

COSMIC NEWSWIRE #26

Warm/Hot Tangible Dark Matter
Challenges Both Intangible Cold Dark Matter
and Modified Gravity (MOG)

LOS ALTOS HILLS, Calif., Nov. 15, 2007 (AScribe Newswire) -- An Oct. 31 article in ScienceNOW Daily News[9], questioning Cold Dark Matter is entitled, "Dark Matter Not a Done Deal?" It refers to the famous August 2006 Bullet Cluster observations, analysis and discovery. ScienceNow says, "A year ago astronomers reported observations that were widely hailed as proof positive for the existence of dark matter."

ScienceNOW then continues with the report of a recent attack on dark matter, as follows: "Not so fast, says astrophysicist John Moffat of the Perimeter Institute for Theoretical Physics in Waterloo, Canada. In a paper, "The Bullet Cluster 1E0657-558 evidence shows modified gravity in the absence of dark matter"[33], published online 25 October in the *Monthly Notices of the Royal Astronomical Society,* Moffat and co-author and Perimeter Institute colleague Joel Brownstein argue that their modified theory of gravity, which they call MOG, can also explain the Bullet cluster discrepancy."

"Einstein argued that gravity arises because mass warps space and time — he even came up with an equation for how the warping works. Moffat says he added 'minimal' additional terms to Einstein's equation that subtly change how gravity behaves on galactic scales. The upshot is that gravity is stronger at these scales than Einstein predicted and that MOG can explain the gravity of the Bullet cluster without dark matter."

This attack on Cold Dark Matter and Bullet cluster conclusions came about because both Cold Dark Matter and the modified theory of gravity called MOG are described only in terms of their gravitational effects. Until the intangible Cold Dark Matter exhibits additional relevant characteristics and parameters, its proof of existence will remain inadequate for its use in modern Big Bang cosmology.

The article "Modern Cosmology: Science or Folk Tale?"[32] in the September-October 2007 issue of American Scientist magazine (which is discussed in Chapter 26 entitled, "British Professor Elegantly Questions Validity of Cold Dark Matter Hypothesis") points out that modern Big Bang cosmology based upon Cold Dark Matter relies upon too few astronomical observations and upon too many hypotheses to be considered *science*.

The problem with the 23-year-old Cold Dark Matter hypotheses is that if we attempt to add another astronomical observation to improve the net significance of the Big Bang cosmology results, we most likely would have to add another hypothesis to the list of Cold Dark Matter hypotheses for the group of them to be compatible with the new astronomical observation. Thus, there would be no improvement in net significance even though there is no shortage of available relevant astronomical observations.

The two previous paragraphs are an attempt to describe the dark matter conundrum posed in "Modern Cosmology: Science or Folk Tale?"[32] by Professor Michael J. Disney of Cardiff University. A response is provided here to this dark matter conundrum, by providing examples of relevant dark matter characteristics and parameters that appear to be fundamental for dark matter, to be used in modern Big Bang cosmology. Since there is not enough known about the intangible Cold Dark Matter to discuss its characteristics, parameters, or its compatibility with the relevant astronomical observations required for modern Big Bang cosmology, a recently discovered tangible, well-characterized dark matter candidate, will be used for these purposes.

The dark matter candidate that will be used is the first and only tangible dark matter candidate that has been discovered to date. It is the warm/hot Relativistic-Proton Dark Matter, which is described as comprising galaxy-orbiting relativistic protons accompanied by relativistic helium nuclei in a nuclei ratio in the range of about 10:1 to 12:1. For narrative simplicity, only the protons will be mentioned from this point forward.

This warm/hot tangible dark matter was first announced and described Dec. 2003 in *How Dark Matter Created Dark Energy and the Sun* and later described in great detail in a second book in May 2006 entitled *Comprehending and Decoding the Cosmos.*

The equivalent proton energies of this type of dark matter range from a relativistic mass of more than 11 times to less than 1100 million times the mass of a proton at rest. When these protons are disturbed from their dark matter halo orbits, they may leave and pass through space as a straggler until they enter a star system, such as our solar system, and collide with atoms, molecules, and particles. In such case the collision will lead to the creation of multitudinous pions that decay into muons, which in turn decay into electrons. These electrons can combine with protons to form hydrogen and with helium nuclei to form helium.

In 1956, Luis W. Alvarez, the 1968 Nobel Prize winner in physics, and his associates, discovered the hydrogen-fusion-catalytic capability of the muon (mu-meson) in a hydrogen bubble chamber. They found that muons, a particle produced in abundance by cosmic rays, catalyzed hydrogen fusion at a low rate at room temperature. The book, *Comprehending and Decoding the Cosmos*[8] (Chapters 2006-26 and 2006-27), posits that muons play a key role in igniting, catalyzing and maintaining hydrogen fusion reactions in stars.

Also, as the relativistic protons move across magnetic field lines in space they will generate synchrotron emission which could be in the infrared, UV, EUV, or x-ray wave-lengths depending upon the proton energies and transverse magnetic field strength. When involved with braking collisions, the protons will emit bremsstrahlung radiation. When they bombard the Earth's atmosphere they are called cosmic ray protons, which were discovered in 1912 by Victor F. Hess, who won the 1936 Nobel Prize in physics for the discovery.

Chapter 27 entitled "Thirteen Astronomical Observations to Solve British Professor's Dark Matter Conundrum" describes 13 astronomical observations fundamental to modern Big Bang cosmology. The description of the characteristics and parameters of Relativistic-Proton Dark Matter in the previous four paragraphs and in the cited references may be utilized to determine its compatibility with each one of the 13 astronomical observations as follows:

Testing 13 Independent Astronomical Observations Fundamental to Big Bang Cosmology for Compatibility with Relativistic-Proton Dark Matter (R-PDM)

1. Hydrogen and helium in an atomic ratio of about 12:1 comprise over 99 percent of the mass of the universe. COMPATIBLE with R-PDM: See page 30 of Book A. (See footnote)

2. Relativistic protons and relativistic helium nuclei, which are common in the universe, are in a nuclei ratio estimated at about 10:1 near Earth. COMPATIBLE with R-PDM: See page 30 of Book A. (See footnote)

3. Dark matter represents about 83 percent of the mass of the universe. COMPATIBILITY with R-PDM is possible.

4. Astronomical data indicate the dark matter particle mass is at least 35 times greater than the mass of a proton at rest. COMPATIBLE with R-PDM: See page 16 of Book A. (See footnote)

5. Dark matter in early epochs was concentrated in long, large, curved filaments, announced by NASA *[18]* on September 8, 2004, which formed galaxy clusters where two dark matter filaments intersected. COMPATIBLE with R-PDM: See Chapter 25.

6. The separation velocities between galaxy clusters began to accelerate about 5 billion years ago, whereas no

acceleration of separation velocities between galaxies or between stars has been reported. COMPATIBLE with R-PDM: See Chapters 18 and 21.

7. Strongly self-interacting dark matter particles are normally in halos of dwarf and low surface brightness (LSB) galaxies while weaker self-interacting dark matter particles are normally located within galaxy clusters orbiting groups of galaxies, implying the existence of two types, forms, or modes of dark matter particles. COMPATIBLE with R-PDM: See Chapter 2006-50 of Appendix C.

8. Starburst galaxies created by the merging of spiral galaxy clusters exhibit extremely high star formation rates. COMPATIBLE with R-PDM: See Chapters 2006-46 and 2006-47 of Appendix C.

9. The blue and blue-white stars, as young as one million years old, are located in the spiral arms of mature spiral galaxies, which also contain five-billion-year-old red stars in their nuclei. COMPATIBLE with R-PDM: See Chapter 2006-48 of Appendix C.

10. The Big Bang, as a thermodynamic process, satisfied the Second Law of Thermodynamics, which requires that the total amount of disorder, or entropy, in the universe, always increases with time. Also, since the Big Bang represents the beginning of time for the universe its entropy should have been at its lowest level for all time. COMPATIBLE with R-PDM: See J. Drexler, 2007, arXiv: physics/0702132 v1 entitled, "A Relativistic-Proton Dark Matter Would Be Evidence The Big Bang Probably Satisfied The Second Law of Thermodynamics"[13].

11. Nuclei [cores] of starburst galaxies usually exhibit new blue star formation while spiral galaxies exhibit new blue star formation in their spiral arms. COMPATIBLE with R-PDM: See Chapters 2006-46 and 2006-48 of Appendix C.

12. Large and mature spiral galaxies were in existence less than 2.5 billion years after the Big Bang. COMPATIBLE with R-PDM: See Chapters 34 and 51 of Book B. (See footnote)

13. Authors of "Missing Mass in Collisional Debris from Galaxies"[27] conclude, "it more likely indicates that a substantial amount of dark matter resides within the disks of spiral galaxies. The most natural candidate is molecular hydrogen in some hard-to-trace form." COMPATIBLE with R-PDM: See Chapter 20[27].

Footnotes to Identify Book A and Book B

Book A: J. Drexler, *How Dark Matter Created Dark Energy and the Sun: An Astrophysics Detective Story*[5] (Universal Publishers, Florida, USA, 2003)

Book B: J. Drexler, *Comprehending and Decoding the Cosmos: Discovering Solutions to Over a Dozen Cosmic Mysteries by Utilizing Dark Matter Relationism, Cosmology, and Astrophysics*[8] (Universal Publishers, Florida, USA, 2006) Seven chapters of this book are in Appendix C.

Drexler first published his basic Relativistic-Proton Dark Matter theory in a December 2003 book[5], his advanced version of the R-PDM theory, which entered into cosmology, in a May 2006 book[8] and his postmodern Big Bang cosmology theory in a February 2007 scientific paper[13].

CHAPTER 29

COSMIC NEWSWIRE #27

Auger Collaboration Probably Detected
Big-Bang-Created UHECR Protons
After Their Ejection by Merging Galaxy Clusters

LOS ALTOS HILLS, Calif., Nov. 26, 2007 (AScribe Newswire) -- The Pierre Auger collaboration, an international project involving 370 scientists and engineers from 17 countries, announced on Nov. 8 the significant discovery of 27 ultra-high-energy cosmic ray (UHECR) protons emitted from some unknown extragalactic sources located within 3 degrees of "active galactic nuclei" (AGNs) within about 250 million light-years of Earth.

There was no explanation of the sources and enormous energies of these UHECR relativistic protons or of their acceleration means.

The following theory may explain the discovery: The Second Law of Thermodynamics required that the Big Bang, while creating the universe's mass and energy, generate primarily relativistic protons in order to minimize the entropy of the universe at the beginning of time. From the generally accepted Big Bang temperatures, some proton energies might have been at energy levels between 1 million EeV to 10 million EeV. Over the subsequent 13.7 billion years the so-called GZK proton-energy-loss effect (explained later) could diminish these proton energies by a factor of 100,000, yet would still permit the arrival at Earth of a small percentage of 10 EeV (10 exa-electronvolt) to 100 EeV ultra-high-energy cosmic ray protons, which could be observed today.

The author presents data and arguments that the recent Auger collaboration discovery of 60 EeV cosmic ray protons from sources "that lie within roughly 250 million light-years of Earth" probably represents these energy diminished Big Bang relativistic protons that were ejected from their long-term steady state orbital paths, around several to tens of spiral galaxies, by transient magnetic field shocks brought about by the merging of galaxy clusters.

The understanding and explanations for both the nature of the sources of the UHECR protons observed by the Auger collaboration and the extremely high energy levels of these UHECRs evolved from over five years of researching UHECRs and Relativistic-Proton Dark Matter by the author. That effort led to scientific books published in 2003[5] and 2006[8] and scientific papers published online on the physics arXiv in 2005[12] and 2007[13].

The Nov. 8 University of Chicago news release contained the following four relevant quotations regarding the discovery of the Auger collaboration:

> "The international Auger collaboration has traced the rain of high-energy cosmic rays [UHECRs] that continually pelts the Earth to the cores of nearby galaxies, which emit prodigious quantities of energy."

> "In the next few years, our data will permit us to identify the exact sources of these cosmic rays and how they accelerate these particles."

> "Cosmic rays — mostly protons — fly through the universe at nearly the speed of light. The most powerful cosmic rays contain more than one hundred million times more energy than the particles produced in the world's most powerful particle accelerator."

> "Scientists have long considered Active Galactic Nuclei (AGN) to be possible sources of high-energy cosmic rays. And while they have now found a strong correlation between the two, exactly what accelerates cosmic rays to such extreme energies remains unknown."

On Nov. 9, 2007, Science magazine published a scientific paper entitled, "Correlation of the Highest-Energy Cosmic Rays with Nearby Extragalactic Objects"[34].

The Nov. 9 Science magazine also published an article by Adrian Cho that contained the following four substantive quotations:

"The cosmic rays do not point precisely to the AGNs; presumably, our galaxy's magnetic field deflects them in transit. Details of the analysis suggest that the cosmic rays are protons."

"Physicists measure the energy of the highest energy rays in exa-electron volts EeV. The Auger team finds that rays with energies higher than 57 EeV — of which they see 27 — generally come from directions within 3° of "active galactic nuclei" (AGNs) that lie within roughly 250 million light-years of Earth."

"Meanwhile, theorists have a puzzle to solve: Exactly how might an AGN accelerate a proton to such mind-boggling energies?"

Reflecting Alan Watson's key comments, Adrian Cho wrote, "The results don't prove AGNs are sources of the rays. Anything else that's distributed on the sky in the same way as AGNs could be the source," Alan Watson says. He added, "For example, galaxies tend to clump, so some other sort of galaxy might be the culprit." (Alan Watson is a professor at the University of Leeds and co-founder of the Pierre Auger Observatory.)

A Nov. 9 ScienceDaily article added, "Galaxies that have an AGN seem to be those that suffered a collision with another galaxy or some other massive disruption in the last few hundred million years."

Google News Nov. 9: Comment by Dr. Paul M. Mantsch, Senior physicist at Fermilab and project manager of the

Pierre Auger Project. A two-paragraph article included the key sentence, "Although violent AGN are good candidates for sources, they might only be tracers for some other kind of sources nearby."

The mysteries of the Auger collaboration discovery reminded Silicon Valley's inventor/scientist, Jerome Drexler, that he had posited a solution to a similar mystery in the summer of 2005 and later wrote about it in Chapter 2006-47 of his May 2006 book *Comprehending and Decoding the Cosmos*[8]. Excerpts from that chapter, as follows, attempt to explain the Auger collaboration's significant discovery of UHECRs from the directions of AGNs without evoking black holes or AGNs:

> Then, on July 29, 2005, Elena Pierpaoli and Glennys Farrar posted a paper on the Physics arXiv, astro-ph/0507679 entitled, 'Massive galaxy clusters and the origin of Ultra High Energy Cosmic Rays', in which the massive galaxy clusters are described as a merging pair of clusters. In their paper, Pierpaoli and Farrar suggest a possible explanation for the observed phenomenon as follows:

> A merging pair of clusters would be expected to have very large scale, strong magnetic shocks which could be responsible for accelerating UHECR even if there is no AGN (active galactic nuclei) or GRB (gamma ray burst) associated with the galaxy clusters.

Note that Pierpaoli and Farrar believe that lower-energy cosmic ray protons are accelerated into UHECRs through magnetic shocks created in the merging galaxy clusters.

Perhaps both research groups are correct in concluding that the UHECRs may have been accelerated. However, there is another possibility. During the pre-merger period, UHECRs, defined as having energies at or above 1 EeV (1 exa-electron volt), might have been orbiting galaxy clusters within their dark matter halos in a steady-state manner according to the Larmor Radius equation. Given the general size of galaxy clusters and the generally accepted magnitude range of the extragalactic magnetic field, one would conclude that most of the pre-merger orbiting protons in the dark matter halos around the galaxy clusters would be UHECRs. [See Appendix B, pages 213-215]

The galaxy cluster merging process would upset the steady-state Larmor orbiting symmetry of the UHECRs. The combining of the magnetic fields of the two merging spiral galaxy clusters could create transient magnetic field distortions (shocks), which would cause a number of UHECRs to be deflected off into space, with some being Earthbound. This theory might be called the deflection-from-orbit theory of UHECR emission. It is presented as a plausible alternative theory to the shock acceleration UHECR theory, which remains unproven according to the two research groups.

The Pierpaoli-Farrar paper indicates that the authors have found data about several UHECR events with energies at about 50 EeV departing from a merging pair of galaxy clusters observed in the SSDS DR3. In the 22 Nov 2005 (v3) abstract of the Pierpaoli-Farrar paper they say, "For cosmic rays with energies above 50 EeV the observed correlation is the strongest for angles of 1.2-1.6 degrees where it has a chance probability of about 0.1 percent."

Also in the v3 version of the paper, the Discussion and Conclusions section says, "Therefore we conclude that the correlations between AGASA UHECR and galaxy clusters do not seem to be driven by the presence of BL Lac or AGN within the galaxy clusters."

Presented here in the previous four paragraphs are (1) Drexler's posited deflection-from-orbit theory of UHECR emission (2) the theory's brief description based upon the Relativistic-Proton Dark Matter hypothesis, and (3) its experimental basis derived from the Pierpaoli-Farrar astronomical data and their conclusions. The task left to complete is the positing of how the relativistic protons orbiting groups of galaxies within a galaxy cluster obtained their energy levels above 50 EeV. The following logical steps should lead toward that result.

Let us now compare the probability of validity of Drexler's Relativistic-Proton Dark Matter hypothesis to the probability of validity of the hypothesis that an AGN mechanism can accelerate protons to energies 100 million times higher than the most powerful particle accelerator on Earth. Before we proceed with that task, let us think about the Pierpaoli-Farrar researchers who in 2005 discovered UHECRs emanating from merging galaxy clusters which do no not exhibit AGNs.

If it required a black hole-AGN accelerator mechanism to create the UHECRs for the Auger collaboration discovery, what other accelerator mechanism was used by the merging galaxy clusters reported by Pierpaoli-Farrar which did not possess AGNs? It is highly unlikely that two completely different accelerator mechanisms exist in the universe that could achieve the enormous proton accelerations described in the previous paragraph. In fact it is highly unlikely that any cosmological accelerator mechanism could achieve those proton accelerations except the Big Bang.

Drexler's scientific books and papers explain that the Big Bang created relativistic protons having energies up to 10 million times higher than 1 EeV UHECRs of today in order for the Big Bang to satisfy the Second Law of Thermodynamics. See Drexler's online published paper, physics/0702132[13], (or Chapter 12) and his Dec.15, 2003 book, which state that the Big Bang generated protons at 10 million EeV. Then over the following 13.7 billion years their energies could decline by the so-called GZK loss mechanism by a factor of 100,000 and still be at the 100 EeV UHECR energy level observed on Earth on rare occasions.

It should be noted that the GZK proton-energy-loss effect pertains to energy losses from pion production by single protons interacting with Cosmic Microwave Background (CMB) photons that limit cosmic ray travel of protons with energies of 60 EeV or more to 300 million light years. (Until recently, 163 million light years was used as the limit.) In that situation the effective CMB photon density far exceeds the proton density. However, when high-flow-level proton streams are orbiting spiral galaxies in the dark matter halos the outer layer protons can shield the inner proton flows from the CMB photons and from the GZK loss effect. This should occur when the effective proton density is at least a large percentage of the effective CMB photon density. (See Appendix A regarding the GZK effect.)

Also, the terms "GZK cutoff" and "GZK limit" are misleading in themselves. The terms imply 60 EeV UHECRs are capable of traveling only about 300 million light years through space, but in actuality that *limit* probably represents only a very large percentage decline for the very highest energy UHECRs and a very much smaller percentage decline for the lowest energy UHECRs.

When the reader becomes convinced that the Relativistic-Proton Dark Matter is a tangible concept and the Big Bang satisfied the Second Law of Thermodynamics through a *violent radial dispersion of relativistic baryons*, then the deflection-from-orbit theory of UHECR emission, posited by Drexler above and in Chapter 2006-47 of his May 2006 book, can be applied to the Auger collaboration discovery.

Note that a corollary emanates from the above presentation that the Auger collaboration discovery itself may represent new and important evidence supporting the validity of Relativistic-Proton Dark Matter and/or the Relativistic Big Bang.

Jerome Drexler sincerely thanks Professor Alan Watson and Dr. Paul Mantsch for their insightful, courageous, and very helpful caveats regarding the involvement of AGNs in the UHECR emission.

Drexler followed his 2003 book[5] with the publishing online of a 19-page scientific paper on April 22, 2005, on the physics website arXiv.org as e-Print No. astro- ph/0504512, and a five-page scientific paper on February 15, 2007, as e-Print No. physics/0702132 (which links UHECRs to the Big Bang and to the Second Law of Thermodynamics).

His 295-page paperback book sequel, entitled *Compre-hending and Decoding the Cosmos*[8], was published May 22, 2006. It solves 15 to 25 previously unsolved cosmic mysteries. This book is now cataloged and available in over 40 astronomy or physics libraries around the world.

CHAPTER 30

COSMIC NEWSWIRE #28

Postmodern Cosmology, Born via 2003 Book *How Dark Matter Created Dark Energy and the Sun*, Challenges Mainstream Cosmology

LOS ALTOS HILLS, Calif., Dec. 5 (AScribe Newswire) Dec. 15, 2007, will mark the fourth anniversary of the publishing of a *postmodern Big Bang cosmology* or *Postmodern Cosmology* theory in a book entitled *How Dark Matter Created Dark Energy and the Sun*. The original form of the cosmology theory is described in Part X (page 79) under the title "Cosmic-Ray Cosmology: Drexler's Unified Theory of Dark Matter, Accelerating Expansion, and Star Formation".

Postmodern Big Bang cosmology, also known as, *Postmodern Cosmology* is based upon utilizing the tangible Relativistic-Proton Dark Matter, and its links to the *Relativistic Big Bang* and to ultra-high-energy cosmic-ray protons. *Postmodern Cosmology* is able to utilize many more astronomical observations and relies on considerably fewer hypotheses than does today's mainstream Big Bang cosmology, which is based upon the intangible Cold Dark Matter.

The problem with the 23-year-old Cold Dark Matter hypotheses is that if we attempt to add another astronomical observation to improve the net significance of the Big Bang cosmology results, we most probably would have to add another hypothesis to the list of Cold Dark Matter hypotheses for the group of them to be compatible with the new astronomical observation. Thus, there would be no improvement in net significance.

That is, the problem with current Big Bang cosmology is not a shortage of available relevant astronomical observations, but that the Cold Dark Matter hypotheses cannot generally show compatibility with an added independent astronomical observation without adding another new hypothesis. On the other hand, increasing the number of astronomical observations will increase the net significance for *Postmodern Cosmology*. The term *postmodern* implies the use of Relativistic-Proton Dark Matter and the *Relativistic Big Bang*.

This problem with Cold Dark Matter comes about because it is described only in terms of its gravitational effects. There is no other proven physical description to utilize. Until Cold Dark Matter exhibits additional relevant physical characteristics and parameters, its proof of existence will remain inadequate for its use in today's Big Bang cosmology.

When it was discovered that the *Postmodern Cosmology* was able to solve more than 15 unsolved cosmic mysteries, a book sequel was authored by Jerome Drexler, entitled *Comprehending and Decoding the Cosmos*[8], which was published May 22, 2006.

Drexler also published online a scientific paper on the physics archive arXiv in 2007 that was fundamental in the final development of *Postmodern Cosmology*, which is based upon Relativistic-Proton Dark Matter. The Feb. 15, 2007 scientific paper, e-Print No. physics/0702132, describes and explains the strong linkage between the Big Bang theory and the Relativistic-Proton Dark Matter theory via the Second Law of Thermodynamics and the entropy level required by the Big Bang at the beginning of time.

Drexler's *Postmodern Cosmology* may create science-fiction-like news headlines when this book is published.

Examples:

The Big Bang Was A Baryon Accelerator Not a Fireball

Relativistic Dark Matter is Proof of Relativistic Big Bang

Accelerating Expansion of Universe is Caused by Synchrotron Emission from Relativistic Dark Matter

Relativistic Big Bang Explains Cosmic Inflation

Cosmic Inflation Supports a Relativistic Big Bang

Ultra-High-Energy Cosmic Ray Protons Bombarding Earth Were Energized and Launched by the Big Bang

Second Law of Thermodynamics for the Universe Required a Low-Entropy Relativistic-Proton Big Bang

Postmodern Cosmology May be Tested by the US in 2009 and Russia in 2010 by EUV/UV Space Telescopes

Postmodern Cosmology is Simple and Beautiful, but Not Too Simple

Postmodern Cosmology: A Cosmology with the Answers

CHAPTER 31

COSMIC NEWSWIRE #29

Nobel Laureates' Queries Point Toward Drexler's Dark Matter Theory and *Postmodern Cosmology*

LOS ALTOS HILLS, Calif., Dec. 11 (AScribe Newswire) – "How inflation happened a split second after the Big Bang." "Identify the exact sources of these cosmic rays and how they accelerate particles." Two different Nobel Laureates in physics raised these two unrelated queries in cosmology publicly in November and December 2007. The timing of these queries is close to the fourth anniversary of a *postmodern Big Bang cosmology* or *Postmodern Cosmology* theory published Dec. 15, 2003, in a cosmology book *How Dark Matter Created Dark Energy and the Sun* authored by Jerome Drexler.

There is the possibility that plausible answers or helpful responses to both of these queries can be derived from a recent cosmology scientific paper utilizing Drexler's dark matter theory and Big Bang cosmology. The paper was published online on the physics arXiv on Feb. 15, 2007 as e-print No. physics/0702132. It is titled "A Relativistic-Proton

Dark Matter Would Be Evidence the Big Bang Probably Satisfied the Second Law of Thermodynamics".

The paper argues that the Big Bang, which occurred at the beginning of time, must have satisfied the Second Law of Thermodynamics. Thus, immediately after the Big Bang the entropy of the universe would be at the lowest level it would reach throughout all time. Therefore the Big Bang should not be characterized, as it has been for over 40 years, as a chaotic fireball explosion associated with a high level of disorder and high entropy.

The very low entropy could be achieved by the Big Bang firing out, in all directions, high-velocity ultra-high-energy (UHE) relativistic protons and helium nuclei close to the well-known nuclei ratio of 12.1. In other words, the Big Bang could be characterized as a *violent radial dispersion of relativistic baryons.*

A very high percentage of their energies would be available to do work since their entropy, a measure of the percentage of their energy unavailable to do work, would be very low. Such a Big Bang, characterized by a *violent radial dispersion* of UHE relativistic nuclei, would be highly efficient and could create very high usable energy and very low entropy, and might be designated a *Relativistic Big Bang*. This concept is fundamental to Drexler's dark matter theory and his *Postmodern Cosmology.*

The *Relativistic Big Bang* would have the protons and helium nuclei being fired out at near the speed of light in almost a purely radial outward direction for a short-time first phase, followed by a second phase during which the magnetic field deflections and electric-charge repulsion of the particles would impart a transverse motion and angular momentum to the particles, thereby greatly reducing their radial outward velocities.

The *Cosmic Inflation* period could be related to the extremely short-time first phase of almost a purely radial outward motion near the speed of light of the relativistic protons and helium nuclei, beginning a split-second after the Big Bang created them.

Hopefully, this explanation will be considered a plausible answer or helpful response to a Nobel Laureate's December 2007 public query, "How inflation happened a split second after the Big Bang."

Now let us consider the query about cosmic rays, "identify the exact sources of these cosmic rays and how they accelerate these particles." The astronomical data used to arrive at answers to this two-part cosmic-ray query will be taken from the reports of the Pierre Auger collaboration, an international project involving 370 scientists and engineers from 17 countries. They announced on Nov. 8, 2007 the significant discovery of 27 Ultra-High-Energy Cosmic Ray (UHECR) protons with energies higher than 57EeV.

These are extremely rare events requiring a relativistic-proton detection system the size of Rhode Island.

The Second Law of Thermodynamics required that the Big Bang, while creating the universe's mass and energy, generate most of the mass in the form of relativistic protons and helium nuclei in order to minimize the entropy of the universe at the beginning of time.

From the generally accepted Big Bang temperatures, some proton energies might have been at energy levels between 1 million EeV to10 million EeV. Over the subsequent 13.7 billion years the so-called GZK proton-energy-loss effect (explained later) could diminish these proton energies by a factor of 100,000, yet would still permit the arrival at Earth of a small percentage of 10 EeV to 100 EeV UHE cosmic ray protons.

The author, Jerome Drexler, believes that the recent Auger collaboration discovery of the higher-than 57 EeV cosmic ray protons probably represents the energy-diminished Big Bang relativistic protons that over billions of years as stragglers in space finally found a home orbiting several to tens of galaxies. Later, they were deflected and ejected from their long-term steady-state orbital paths, around several to tens of spiral galaxies, by transient magnetic field shocks brought about by the merging of two galaxy clusters.

It should be noted that the GZK proton-energy-loss effect pertains to energy losses from pion production by single protons interacting with Cosmic Microwave Background (CMB) photons that limit cosmic ray travel, of protons with energies of 60 EeV or more, to 300 million light years. In this situation, the effective CMB photon density far exceeds the proton density.

However, when high-flow-level proton streams are orbiting groups of spiral galaxies in dark matter halos the outer layer protons can shield the inner proton flows from the CMB photons and thus from the GZK loss effect. This weakening of the GZK loss effect should occur when the effective proton density is a substantial fraction of the effective CMB photon density. (See Appendix A regarding the GZK effect.)

Also, the terms GZK cutoff and GZK limit are misleading. The terms imply 60 EeV UHECRs are capable of traveling only about 300 million light years through space, but in actuality that *limit* probably represents only a very large percentage decline for the very highest energy UHECRs and a very much smaller percentage decline for the lowest energy UHECRs.

Hopefully, this explanation will be considered a plausible answer or helpful response to a Nobel Laureate's November 2007 public query, "identify the exact sources of these cosmic rays and how they accelerate these particles."

For a more information about sources of UHE cosmic ray protons and their acceleration means, see reference[34] and Chapter 29 of this book and Chapter 2006-47 of *Comprehending and Decoding the Cosmos*[8] in Appendix C.

CHAPTER 32

COSMIC NEWSWIRE #30

Possibly Solving the *Missing Baryon* Mystery

*Cosmic newswire title: Drexler's 'Missing-Baryons
Dark Matter' May be a Solution to
a NASA Major Puzzle: the 'Missing Baryons'*

LOS ALTOS HILLS, Calif., Jan. 17 (AScribe Newswire) – A NASA Web site says the location of the "missing baryons" is a "major astronomical puzzle". Science magazine published a special issue on January 4 focusing on the universe's *missing baryons* and the *cosmic web*. The issue features an article, "Missing Baryons and the Warm-Hot Intergalactic Medium"[45] by Harvard-Smithsonian scientists.

An introduction to the *missing baryon* mystery contains these excerpts: "The exact nature of the dark matter that makes up 95% of the *cosmic web* remains baffling, but things aren't much better for the remaining 5% that we can see. These are the baryons — protons and neutrons — that make up ordinary matter, yet we can account for only about half of the baryon mass predicted by the standard cosmological model. The missing baryons might be lurking in the cosmic web…"

Let us proceed on the possibility that the *missing baryons* have not been found because they are protons and helium nuclei moving at relativistic velocities. This approach is hopeful because Relativistic-Baryon Dark Matter requires that a small mass of baryons — protons, neutrons, and helium nuclei — be an integral part of this dark matter, separated from the hydrogen and helium of the universe until the dark matter baryons eventually slow down.

Jerome Drexler discovered Relativistic-Baryon Dark Matter, also known as Relativistic-Proton Dark Matter, six years ago. This later led to his February 15, 2007 scientific paper published online on the physics arXiv entitled "A Relativistic-Proton Dark Matter Would Be Evidence the Big Bang Probably Satisfied the Second Law of Thermodynamics". Basically, the paper argues that the Big Bang, which occurred at the beginning of time, must have satisfied the Second Law of Thermodynamics. Thus, immediately after the Big Bang the entropy of the universe would be at the lowest level it would reach throughout all time.

According to the scientific paper, the very low entropy could be achieved by the Big Bang firing out, in all directions, high-velocity ultra-high-energy (UHE) relativistic protons and helium nuclei close to the well-known nuclei ratio of

12:1. In other words, the Big Bang could be characterized as a *violent radial dispersion of relativistic baryons.*

A very high percentage of their energies would be available to do work since their entropy, a measure of the percentage of their energy unavailable to do work, would be very low. Such a Big Bang, characterized by a *violent radial dispersion* of UHE relativistic protons and helium nuclei, would be highly efficient and could create very high usable energy and very low entropy. It could be designated a *Relativistic Big Bang.*

About 83 percent of the *Relativistic Big Bang's* mass output would represent the universe's dark matter today, which should have a proton/helium nuclei ratio in the range of about 10:1 to 12:1. It has been designated Relativistic-Baryon Dark Matter, but could have been named *Missing-Baryons Dark Matter.*

Drexler has provided overwhelming evidence supporting his Relativistic-Baryon Dark Matter theory in astro-cosmology books published in 2003 and 2006 and in scientific papers published online on the physics arXiv in 2005 and 2007. This same dark matter theory has proven itself over and over again during the past three years in providing published plausible solutions to over two-dozen cosmic mysteries.

For examples, on December 11, 2007, Drexler published online a plausible solution to a fundamental mystery involving *cosmic inflation.* On November 26, 2007 he published plausible solutions online to mysteries involving the departing locations and the energy source for the ultra-high-energy cosmic ray protons detected by the Pierre Auger detectors in Argentina.

Considering the above, the Relativistic-Baryon Dark Matter theory could provide a plausible solution for the mystery of the *missing baryons.*

The Science article "Missing Baryons and the Warm-Hot Intergalactic Medium" reads, "Today we can account for less than 50% of the baryon mass [protons and neutrons] predicted by the Standard Cosmological Model (SCM), implying that at least 50% of the baryons are now *missing.*"

The article indicates that 10 billion years ago baryonic matter appeared to total more than twice as much as it does now. The missing portion is normal baryonic matter made up of protons and neutrons, which includes helium nuclei.

In the Relativistic-Baryon Dark Matter theory the dark matter of the universe makes up about 25 percent of the total mass-energy of the universe, matching the SCM estimate. Relativistic-Baryon Dark Matter is comprised of a *baryon mass* of protons and helium nuclei rapidly moving at a velocity near the speed of light. It contains an enormous energy, represented by an order-of-magnitude higher *relativistic mass* than the *baryon mass*. This *relativistic mass* represents the mass-energy equivalence under Einstein's 1905 Special Theory of Relativity.

In summary, the Relativistic-Baryon Dark Matter is comprised of a *baryon mass* speeding at relativistic velocities. This *baryon mass* could represent the entire universe's *missing baryons* or a portion. Perhaps Harvard-Smithsonian, NASA, and Science magazine should look into this research approach to solving the "major astronomical puzzle" of the *missing baryons* as well as the major mystery of dark matter itself.

CHAPTER 33

COSMIC NEWSWIRE #31

NASA to Probe Dark Matter with EUV/UV

Cosmic newswire title: Dark Matter Radiation,
Predicted in Drexler's 2003 Book, Testable
When NASA Probes Dark Matter Web's EUV and UV

LOS ALTOS HILLS, Calif., Jan. 11 (AScribe Newswire) – On Jan. 8, 2008, NASA announced an August 2008 Hubble servicing mission[46] as follows:

"The new instruments to be installed on the telescope are the [EUV and UV] Cosmic Origins Spectrograph, or COS, and the Wide Field Camera 3, or WFC3. Among its many goals, COS will probe the *cosmic web*. This large-scale structure of the universe has its form determined by the gravity of dark matter and can be traced by galaxies and intergalactic gas. COS also will explore how this web has evolved over billions of years and the role it plays in the formation and evolution of galaxies."

In his Dec.15, 2003 book, *How Dark Matter Created Dark Energy and the Sun,* inventor-scientist Jerome Drexler predicts that dark matter is a source of synchrotron radiation. With the broad range of proton energies in Relativistic-Baryon Dark Matter, physics textbooks calculate that synchrotron radiation, from this dark matter, would include extreme ultraviolet (EUV) and UV photon emission.

Drexler's 2003 book also states that synchrotron radiation from dark matter may explain the accelerating expansion of the universe, announced by Saul Perlmutter and Brian Schmidt in 1998. His 2003 accelerating-expansion theory is that synchrotron radiation from the dark matter of galaxy clusters would continually lower the clusters' relativistic mass and thus "[r]educe each galaxy cluster's gravitational attraction to nearby galaxy clusters, thereby facilitating their more rapid separation" and "[r]aise the galaxy clusters' (separation) velocities under the Law of Conservation of Linear Momentum." Aspects of this quadruple-physics-law acceleration theory should be testable by NASA.

Science magazine published a special issue on January 4, 2008 focusing on the universe's *cosmic web* the filamentary dark matter structure that appears to connect all galaxy groups and galaxy clusters. Adrian Cho's Science article "Untangling the Celestial Strings"[44] begins:

> "Unraveling that *cosmic web* presents astronomers, astrophysicists, and cosmologists with their next great challenge. The (cosmic) web is the framework on which the universe is built. It consists primarily of *dark matter*, mysterious stuff that makes up 85% of the matter in the universe but has revealed itself only through its gravity. Enormous filaments and blobs of

the stuff condensed as the universe matured. Within them nestle the galaxies and their stars, creating streams of light stretching between inky voids."

Scientists have been searching for every possible clue to the nature of dark matter; and the *cosmic web* may provide some of them. The *cosmic web* structure may also provide clues as to whether the top-down theory of galaxy formation is more applicable to most of the universe than the bottom-up theory. Every piece we manage to fit into a jigsaw puzzle makes the next piece a little easier to discover. If the *cosmic web* facilitates the discovery of the nature of dark matter or the correct galaxy formation theory, the discovery of the other could follow promptly.

The Science article, "The Cosmic Web in Our Own Backyard"[47], begins with:

"On the largest scales, matter is strung out on an intricate pattern known as the cosmic web. The tendrils of this web should reach right into our own cosmic backyard, lacing the Galactic halo with lumps of dark matter. The search for these lumps, lit up by stars that formed within them, is a major astronomical endeavor..."

The *cosmic web* was discovered in the fall of 2004 in the Fornax galaxy cluster. A September 9, 2004 news release from NASA (and Harvard) entitled "Motions in nearby galaxy cluster reveal presence of hidden superstructure"[18], regarding Chandra x-ray images of the Fornax cluster, states:

> "Astronomers think that most of the matter in the universe is concentrated in long large filaments of dark matter and that galaxy clusters are formed where these filaments intersect." The researchers' paper (astro-ph/0406216) is entitled, "The Chandra Fornax Survey - I: The Cluster Environment"[42].

This astronomically discovered filamentary description of dark matter appears to be much more compatible with the Relativistic-Baryon Dark Matter theory than with the non-baryonic Cold Dark Matter theory. It seems highly unlikely that the filamentary structure of dark matter of the *cosmic web* could be created by very slow moving, weakly interacting (only through gravitational tidal forces) particles.

The vision of large, long dark matter filaments crisscrossing the cosmos gives the impression of high-velocity particles. The crashing of intersecting dark matter filaments creating galaxy clusters gives the impression of a top-down theory of galaxy formation. Both of these impressions point toward and lend support to the Relativistic-Baryon Dark Matter theory and cosmology.

Furthermore, the theoretical weakly interacting massive particles (WIMPs) of Cold Dark Matter, being non-baryonic, cannot be transformed into hydrogen and helium, the substance of galaxies (and stars), where the filaments of dark matter intersect. On the other hand, the relativistic baryons (protons and helium nuclei), of Relativistic-Baryon Dark Matter, can provide the necessary hydrogen and helium to create the galaxy clusters and galaxies where the long filaments of dark matter intersect.

For the above reasons, the September 2004 reports of cosmic filamentary dark matter and the January 2008 reports of *cosmic web* dark matter seem to be very supportive of Drexler's Relativistic-Baryon Dark Matter theory.

CHAPTER 34

The Concluding Chapter

Weaving Together Seven Cosmic Phenomena to Form *Postmodern Cosmology* Fabric

The only publicly announced plausible explanation for the *cosmic web* is disclosed in Chapter 33. It requires that dark matter be comprised of ultra-high-energy relativistic protons.

The only publicly announced plausible explanation for *cosmic inflation* is disclosed in Chapter 31. It requires that dark matter be comprised of ultra-high-energy relativistic protons.

The only publicly announced plausible explanation for ultra-high-energy *cosmic ray protons* with energies above 60 EeV bombarding Earth's atmosphere is disclosed in Chapter 29. It requires that dark matter be comprised of ultra-high-energy relativistic protons.

The only plausible explanation for the *accelerating expansion of the universe*, publicly announced to date, is disclosed in Chapter 21. It requires that the dark matter of the universe be comprised of ultra-high-energy relativistic protons.

The only plausible explanation for the *Big Bang satisfying the Second Law of Thermodynamics*, publicly announced to date, is disclosed in Chapter 12. It requires that dark matter be comprised of ultra-high-energy relativistic protons.

Perhaps the dark matter of the universe is indeed comprised of ultra-high-energy relativistic protons.

Postmodern Cosmology or Postmodern Big Bang Cosmology is a cosmology, discovered and developed by Jerome Drexler during the period 2002 to 2007, in which the output of the *Big Bang* was comprised almost entirely of relativistic protons and helium nuclei in a nuclei ratio between 10:1 and 12:1 and *dark matter* is comprised of the same relativistic subatomic particles in a somewhat similar ratio.

Thus, it appears that the Relativistic-Baryon Dark Matter was created in its entirety by the Big Bang. For an understanding of how and why this happened see Chapters 10, 12, 13, 29, and 31. Additional discussion of dark matter and the Big Bang can be found at the end of this chapter.

Cosmic Inflation is the hyper-rapid, extremely short, speed-of-light growth period of the universe that followed the Big Bang by a fraction of a second. According to scientists, "the particle physics mechanism responsible for cosmic inflation is not known." By using publicly available data, the previous chapters, and the laws of physics, Drexler has derived a plausible explanation for the nature of *cosmic inflation* in Chapter 31.

From the previous paragraphs and the generally accepted temperature of the Big Bang, the *ultra-high-energy cosmic ray protons and helium nuclei* that bombard the Earth's atmosphere every day appear to have been created by the Big Bang. What is the difference between these Big Bang-created relativistic subatomic cosmic ray particles and the relativistic subatomic dark matter particles? Answer: There is no difference; they are one and the same.

Eventually, these subatomic relativistic particles will form the dark matter halos around spiral galaxies and more energetic dark matter halos around groups of galaxies within galaxy clusters, probably through *astrophysical emergence* as explained in Chapter 2006-44 of the May 2006 book[8].

The magnetic shocks, caused by the merging of galaxy clusters, experienced by the orbiting relativistic protons would cause some of them to be deflected and ejected from their galaxy cluster orbits as explained in Chapters 2006-47 and 2006-45 of the May 2006 book and Chapter 29 of this book. A very small percent of these proton stragglers bombard the Earth's atmosphere. We earthlings call these stragglers from dark matter *ultra-high-energy (UHE) cosmic ray protons* or UHECRs.

Drexler's discovery that the UHE cosmic ray protons derived their energy from the Big Bang is explained in the six-page Part IV of the December 2003 book *[5]*. Part IV is titled, "Are UHE Dark Matter Halo Protons Relics of the Big Bang?" More current explanations of this phenomenon and/or Drexler's discovery of the protons' departing locations are found in Chapters 2006-45 and 2006-42 of the 2006 book *[8]* and/or in Chapters 29 and 31 of this book.

The terms *dark energy* and a*ccelerating cosmos* are linked. The *accelerating cosmos,* discovered in 1998, is attributed by scientists, to a mysterious *dark energy.* The expanding cosmos is clearly accelerating, but the exact nature of *dark energy* has remained a mystery. In the case of the accelerating cosmos, Drexler's discovery is in the form of a plausible explanation of the physics of the phenomenon, which will be mentioned after *dark energy* is discussed.

Dark energy is clearly selective with regard to the celestial bodies that it affects or to the degree it affects large or small celestial bodies. It causes an acceleration of the separation velocities between galaxy clusters, but does *not* cause an acceleration of the separation velocities between galaxies or a velocity acceleration between stars. A gravitational field does not have different laws applying to different sized or shaped objects. This led Drexler to consider whether a *dark energy field,* favored by many cosmologists, is an erroneous concept.

However, Drexler has discovered a plausible explanation for a form of *dark energy* without using a *dark energy field* concept. Chapter 21 explains why this *dark energy* could cause an acceleration of the separation velocities between galaxy clusters, but would not cause an acceleration of the separation velocities between galaxies or separation-velocity acceleration between stars. This is an important distinction based upon astronomical observations.

Since this type of *dark energy* effect appears to be caused by different levels of synchrotron emission from the relativistic protons orbiting single galaxies versus those protons orbiting groups of galaxies, this type of *dark energy* possibly could be related to the relativistic mass-energy of the universe's relativistic protons and helium nuclei. See Chapters 2006-17, 2006-24, 2006-25, and 2006-30 of the May 2006 book[8].

Drexler's discovery regarding the *accelerating universe* is found in the December 2003 book in the 13-page Part VIII beginning on page 56 entitled "The Accelerating Expansion of the Universe and Dark Energy". It is also covered in Chapters 2006-17, 2006-24, 2006-25, and 2006-30 of his May 2006 book. In this book it appears in Chapters 18 and 21.

The *cosmic web* is considered to be the framework on which the universe is built. It is comprised primarily of *dark matter* that makes up about 83 percent of the mass of the universe. It is explained in Chapter 33.

Drexler's *dark matter* discovery is that dark matter is comprised of galaxy-orbiting relativistic protons and helium nuclei in a nuclei ratio of between 10:1 (see page 30 of the December 2003 book[5]) and 12:1.This was first published in the December 2003 book. It is also discussed in great detail in the May 2006 book and in the two scientific papers published online on the physics arXiv on April 22, 2005[12] and on February 15, 2007[13].

Drexler's *Big Bang* discovery is that in order for the Big Bang to satisfy the Second Law of Thermodynamics, which required that the universe's entropy would be the lowest for all time, the vast majority of the matter and energy it created would have been primarily in the form of relativistic protons. This discovery is explained in Chapters 12 and 13 of this book and in Drexler's February 15, 2007 paper[13] on the physics arXiv.

Note that all seven cosmic phenomena that form the basis of Drexler's *postmodern cosmology* or *postmodern Big Bang cosmology* appear to be compatible with one another and appear to rely on Relativistic-Baryon Dark Matter, which is a more accurate name than the formerly used, Relativistic-Proton Dark Matter.

At this time, there is no other published cosmology that exhibits a higher degree or level of compatibility with the seven cosmic phenomena. They represent the seven pillars of support for *postmodern cosmology.*

APPENDIX A

The Scientific Community's Long-Held Objections to Any Proton Dark Matter Theory are Summarized by the Author as Follows:

1. Scientists' arguments regarding nucleosynthesis, abundance ratios, and the amount of deuterium are as follows: Deuterium was created only in the very early Universe. Since deuterium is fragile, it is not present in the stars. The more deuterium there is today in the Universe, the fewer baryons there are since baryons convert deuterium. The high abundance of deuterium in the Universe today implies a density of ordinary baryonic matter of between 5% and 10% of the critical mass density of the Universe. The critical mass density is the minimum mass density required if the expansion of the Universe is eventually to cease. Since the scientific community believes that the actual mass density of the Universe is much greater than the 5% to 10% of the critical mass density, scientists believe that dark matter must be something other than baryons.

2. The measured cosmic microwave background (CMB) fluctuations indicate that the mass density ripples were too small in the early Universe to attract baryons and evolve into galaxies. To attract baryons before the mass density ripples would disperse, the CMB fluctuations must be of the order of one part in 100, whereas mass density fluctuations observed by COBE are of the order of one part in 100,000. That also led the astrophysicists

and cosmologists to conclude that the vast majority of the matter in the Universe must be comprised of something other than baryons.

3. Regarding flatness (Euclidean geometry) and the critical mass density, scientists believe that the geometry of the Universe is closer to the ordinary Euclidean geometry than any other geometry. Also, it is known that if the Universe had the critical mass density, the Universe could be described in terms of Euclidean geometry and would therefore be "flat." Scientists believe that if all the luminous (ordinary) matter in the galaxies were dispersed in the Universe, the mass density of the Universe would be many times smaller than the critical mass density. From the speeds of stars around the centers of galaxies and the speeds of galaxies moving within galaxy clusters, it appears that dark matter in the Universe may be about 10 times more massive than luminous ordinary matter and about six times more massive than the total ordinary matter.

Author's Response Regarding the Above Three Objections to the Proton-Dark-Matter Candidacy:

For many years, baryons have been ruled out as a DM candidate because the primordial nucleosynthesis calculations and other cosmological considerations indicate a very low *particle abundance* of baryons in the Universe. However, this argument does not address the large quantity of relativistic protons/baryons in the Universe. For example, a 15% ratio of relativistic protons to non-relativistic baryons in the Universe satisfies the particle-abundance-maximum constraint and also satisfies the mass-abundance-minimum constraint because the DM particles, having a very large relativistic mass, are still able to form 80% to 85% of the mass of the Universe.

It is widely accepted by cosmologists that ordinary matter in the Universe totals about 4% of the total mass/energy and that dark matter totals about 23% of the total mass/energy; therefore, the mass energy of dark matter in the Universe is about six times higher than that of ordinary matter.

This same ratio appears plausible for relativistic proton dark matter. If, for example, the total number of relativistic dark matter protons in the Universe were no more than 15% of the total number of baryons in the Universe, the relativistic proton dark matter theory would be compatible with the proton limitations of each of the above three objections of the scientific community.

That is, if the average relativistic mass of the dark matter protons is, say, 50 times the rest mass of a proton and their number is, say, about 12% of the number of baryons in the Universe, then the dark matter relativistic mass would total about six times that of ordinary matter mass in the Universe. This matches cosmologists' estimates mentioned above. For an average relativistic mass ratio of 50, the average proton energy would be about 5×10^{10} eV, which is near the low end of the cosmic ray proton energy range of about 1×10^{10} eV to 5×10^{19} eV (see Appendix B, Slide #17).

The Objection of Some Scientists is Based on the GZK (Greisen-Zatsepin-Kuzmin) Cut-Off Theory Argument:

In the past, the 1966 GZK cut-off theory was sometimes used as one of the arguments to counter the relativistic proton dark matter theory. The GZK cut-off theory predicts that because of the interaction with the cosmic microwave background, relativistic protons cannot have energies higher than 6×10^{19} eV at the Earth, since above those levels they would lose energy rapidly in collisions with the CMB.

Author's Response to the GZK Argument:

Some astrophysicists who have been inclined to use the 1966 GZK cutoff theory to rule out the relativistic proton dark matter model were apparently not aware of the 1998 paper designated hep-ph/9808446 and entitled "Evading the GZK Cosmic-ray Cutoff"[56] by Sidney Coleman and Nobelist Sheldon L. Glashow or the 1997 published chapter of Nobelist James W. Cronin in *Unsolved Problems in Astrophysics*[57]. James Cronin's chapter in the book reported two observed cosmic ray protons with energies about four times higher than the theoretical GZK proton-energy cutoff.

The presence of UHE cosmic ray protons in the solar system, including a very few with energies well above the GZK cutoff, adds some plausibility to an anti-GZK effect.

Also, researchers have reported an anti-GZK effect that arises when UHE relativistic protons moving through an intergalactic magnetic field experience diffusive propagation. They report that this effect causes a jump-like increase in the distance UHE protons can travel. From its description in the literature, apparently relativistic protons orbiting a galaxy in its halo's magnetic field also would experience an anti-GZK effect. For example, see astro-ph/0507325 authored by R. Aloisio and V. S. Berezinsky[58].

Further, it should be noted that under the relativistic proton dark matter theory/cosmology, the only relativistic protons that could have energies even close to the theoretical GZK cutoff are those orbiting a galaxy supercluster. For this case, the proton flux density would be extremely low (see Appendix B, Slide #17). For example, cosmic ray protons at an energy level of 10^{19} eV or above striking the Earth's atmosphere total only 3 to 4 per square kilometer per century.

APPENDIX B

Presented here are 18 selected pages involving six important references from J. Drexler's book *"How Dark Matter Created Dark Energy And The Sun"* (Universal Publishers Boca Raton, Florida USA, 2003).

Energies of Relativistic Protons Versus Their Relativistic Mass – Where Do They Exist in Nature?

Energy 9.38 x 10^8 eV	Relativistic Mass of a Proton In Terms of Its Rest Mass, m_0 m_0
10^{10} eV	11 m_0
10^{11} eV	110 m_0
10^{12} eV	1,100 m_0
10^{13} eV	11,000 m_0
10^{14} eV	110,000 m_0
10^{15} eV	1,100,000 m_0
10^{16} eV	11,000,000 m_0
10^{17} eV	110,000,000 m_0
10^{18} eV	1,100,000,000 m_0

$$\text{Relativistic Mass} = \frac{\text{Energy (in joules)}}{C^2 \text{ (in meters/sec)}}$$

Page 22 of *How Dark Matter Created Dark Energy And The Sun*

SLIDE #15

Such Highly Energetic Protons Can Be Found Striking the Earth's Atmosphere as Cosmic Ray Protons

Approximate Kinetic Energy	Approximate Cosmic Ray Flux On the Earth's Atmosphere
10^8 to 10^{10} eV	Slightly less than 1,000 particles per square meter per second
10^{11} eV	One particle per square meter per second
7×10^{15} eV	One particle per square meter per year
3×10^{18} eV	One particle per square kilometer per year
10^{19} eV	3 to 4 particles per square kilometer per century

Page 23 of *How Dark Matter Created Dark Energy And The Sun*

SLIDE #16

Cosmic-Ray Energy Distribution
at the Earth*

CERN Courier, Vol. 35, No. 10, December 1999

[See Slide #16 for the Key Data Points]

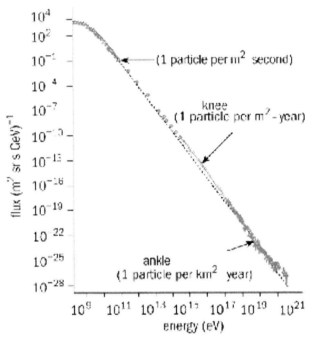

"The cosmic-ray energy distribution shows remarkable uniformity over 10 orders of magnitude. However, there are two kinds. The ACCESS experiment is designed to investigate 'the knee' (near 10^{15} eV)."

*Available on the Internet at http://www.cerncourier.com/main/article/39/10/8/1.

Page 24 of *How Dark Matter Created Dark Energy And The Sun*

SLIDE#17

The Accelerating Expansion Between Galaxy Clusters (Not Between Galaxies)

- Note that Michael Turner points out (see slide #54) that the dark energy "isn't found in galaxies or even clusters of galaxies." This is an extremely important point. This statement means that dark energy pushes galaxy clusters apart but doesn't push galaxies apart within clusters and doesn't push stars apart within a galaxy.

- This is observed locally. While the Universe is expanding, in our Local Group of galaxies Milky Way and Andromeda are moving towards each other at 119 km/sec. Also, our Local Group is moving toward the Local Supercluster at 600km/sec. This illustrates that local strong gravitational forces can overcome the dark energy antigravity forces. This also suggests that perhaps in the earlier, smaller, and denser Universe, when the Universe was less than about 8.2 billion years old, the closeness of the galaxy clusters favored dark matter gravity over dark energy antigravity, and expansion acceleration could not take hold. When the Universe was 8.7 billion years old, the expansion acceleration began, according to Adam Riess. (See slide #53.)

Page 65 of *How Dark Matter Created Dark Energy And The Sun*

SLIDE #58

Jerome Drexler's Theory of the Accelerating Expansion Between Galaxy Clusters

- As the synchrotron radiation emission of gamma rays continues, not only does the kinetic energy of the halo UHE protons fall, but their relativistic mass will fall as well (slides #39 and #40). See slide #15 and assume that the average kinetic energy of the UHE protons declined from 10^{16} eV to 5×10^{15} eV over a period of time in the dark matter halo of some galaxy cluster. This represents a decline in the dark matter mass in the halo of that galaxy cluster of 50% and perhaps a 40% decline in the mass of the combined galaxy cluster and its halo.

- Such a reduction in the dark matter halo mass around galaxy dusters would:

  (1) Raise the galaxy clusters' velocities under the Law of Conservation of Linear Momentum. (See footnote on slide #75.) [Drexler]

  and

  (2) Reduce each galaxy cluster's gravitational attraction to nearby galaxy clusters, thereby facilitating their more rapid separation. [Drexler]

Page 66 of *How Dark Matter Created Dark Energy And The Sun*

SLIDE #59

Drexler's Theory of the
Accelerating Expansion Between
Galaxy Clusters

- The gravitational attraction effect of item (2) on the previous slide will diminish through the years as the nearby galaxy clusters become more distant. Note from slide #54 that from the age of 300,000 years until today, the spacings between galaxies increased by a factor of 1,000.

- In an expanding Universe, all galaxy clusters are moving away from each other. Meanwhile, the masses of their dark matter halos of UHE protons are declining because of the synchrotron radiation energy losses. As a result, the velocity of every galaxy cluster should rise (owing to the reduced gravitational attraction between them and the Law of Conservation of Linear Momentum), thereby accelerating the expansion of the Universe. [Drexler]

Page 67 of *How Dark Matter Created Dark Energy And The Sun*

SLIDE #60

Drexler's Theory of the Accelerating Expansion Between Galaxy Clusters

- The antigravity repulsion between galaxy clusters is proportional to the relative decrease in mass of their dark matter halos of UHE protons owing to the protons' loss of kinetic energy through synchrotron radiation. [Drexler]

- Gravitational attraction is directly proportional to the product of the masses of two galaxy clusters and inversely proportional to the square of the distance between them.

- Thus, for the earlier, smaller, and denser Universe more than five billion years ago, the smaller distances between galaxy clusters caused the gravitational attraction between them to be high because of the inverse square of those smaller distances between galaxy clusters. This inverse-square relationship may be a principal reason that the accelerated expansion did not begin until five billion years ago when the conservation-of-momentum effect and reduced galaxy cluster mass finally overcame the gravitational attraction between galaxy clusters. [Drexler]

Page 68 of *How Dark Matter Created Dark Energy And The Sun*

SLIDE #61

Drexler's Theory:
UHE Cosmic-Ray Nuclei May Have Facilitated the Triggering of the Sun's Fusion Reaction

- With the UHE cosmic ray protons having a kinetic energy in the range of 10^{10} to 10^{20} electron volts, they not only added mass and fuel to the formation of the Sun but considerable nucleus-to-nucleus collision energy as well.

- These UHE nuclei provide a clue that the UHE protons and heavier high-collision-cross-section UHE nuclei may have facilitated the triggering of the Sun's fusion reactions and its birth.

- The traditional theory of the Sun formation involving a hydrogen gas cloud, forces of gravity, compression, and high temperature heating may have to be modified. (For further information about the UHE cosmic-ray nuclei, see slide #17, an energy distribution graph that is a plot of cosmic-ray particle flux versus particle energy.) [Drexler]

Page 78 of *How Dark Matter Created Dark Energy And The Sun*

SLIDE #71

What is the Difference Between a UHE Proton and a Cosmic-Ray Proton Bombarding the Earth?

- When a UHE proton in the halo of the Milky Way (a spiral galaxy) loses a significant portion of its kinetic energy over billions of years through synchrotron radiation, the proton will eventually plummet into the galaxy, thereby accelerating its energy loss. It then becomes one of the cosmic ray protons which bombard wide regions of the galaxy, including the solar system, and may have played a role in creating the Sun and other stars as explained in slides #62 – #71.

- With UHE protons remaining active for billions of years, they may be thought of as "immortal" UHE protons while at end of their life they become "mortal" cosmic ray protons plummeting into the galaxy in what I call a "death spiral."

Page 85 of *How Dark Matter Created Dark Energy And The Sun*

SLIDE #78

The Transformation of "Immortal" UHE Protons into "Mortal" Cosmic-Ray Protons Through the "Death Spiral"

- The synchrotron radiation loss of a relativistic charged particle is inversely proportional to both the radius of curvature of its path and the fourth power of its mass.

- The radius of curvature of a UHE proton's spiral path is equal to the Larmor Radius (see slide #85) and is directly proportional to the kinetic energy of the proton and inversely proportional to the magnetic field strength.

- The extragalactic magnetic field is reported to be about 1×10^{-9} gauss while the magnetic field in the interior of the Milky Way is about 2,000 times greater, at 2×10^{-6} gauss.

- Therefore, in the extragalactic dark matter halo of the galaxy, the magnetic field is very weak, the kinetic energy of the protons is high, the synchrotron radiation losses are extremely low, and the UHE proton may be able to circulate for billions of years. [Drexler]

Page 86 of *How Dark Matter Created Dark Energy And The Sun*

SLIDE #79

The Transformation of "Immortal" UHE Protons into "Mortal" Cosmic-Ray Protons Through the "Death Spiral"

- After billions of years in the extragalactic halo of a spiral galaxy, some of the UHE protons should eventually lose enough energy that their spiral paths will be reduced in diameter and the UHE protons will approach the surface of the galaxy. When the UHE halo protons enter the galaxy, their energy has perhaps diminished by a factor of 10 or so and the magnetic field might be about 100 times greater, thereby increasing the synchrotron radiation loss by a factor of 1,000. (The radius of a proton's spiral path is called the proton's Larmor Radius, which can be calculated as shown in slide #85.)

- Thus, as a UHE proton enters the galaxy, its energy will plummet further, say to half, thereby doubling the synchrotron radiation loss. The proton's energy will drop more and more, and it will enter into the "death spiral" as it plunges deeper into the galaxy and begins to be known as a cosmic-ray proton.

Page 87 of *How Dark Matter Created Dark Energy And The Sun*

SLIDE #80

The Transformation of "Immortal" UHE Protons into "Mortal" Cosmic-Ray Protons Through the "Death Spiral"

- The rate of synchrotron radiation is inversely proportional to the fourth power of the mass of the particle. Therefore, synchrotron radiation from protons is infinitesimal compared to synchrotron radiation from electrons. More precisely, proton synchrotron radiation losses are lower than radiation losses from electrons, following the same radius of curvature path, by a factor of about 11 trillion (the proton/electron mass ratio of 1,836 to the fourth power).

- The discussion on slides #79 and #80 has to do with spiral galaxies, which are known to have a dark matter halo and also a black hole containing a few million solar masses.

Page 88 of *How Dark Matter Created Dark Energy And The Sun*

SLIDE #81

Drexler's Cosmic-Ray Cosmology
Applied to Galaxy Formation

The Proton Larmor Radius

- A proton crossing an orthogonal magnetic field enters into a spiral path. The radius of one cycle of that spiral path is called the Larmor Radius.

- The Larmor Radius of a proton can be calculated as:

$$r = 110 \text{ Kpc} \times \frac{10^{-8} \text{ gauss}}{B} \times \frac{E}{10^{18} \text{ eV}}$$

  where Kpc means kilo parsec and
  one parsec equals 3.26 light-years

- The galactic magnetic field within the Milky Way is approximately 2×10^{-6} gauss compared to the extragalactic magnetic field at 1×10^{-9} gauss.

Page 92 of *How Dark Matter Created Dark Energy And The Sun*

SLIDE #85

Drexler's Cosmic-Ray Cosmology
Applied to Galaxy Formation

The Proton Larmor Radius

- The Larmor Radius for a 10^{16} eV proton in the Milky Way halo's extragalactic magnetic field of 10^{-9} gauss is 11 Kpc; for a 10^{17} eV proton it is 110 Kpc; and for a 10^{18} eV proton it is 1,100 Kpc.

- The diameter of the Milky Way galaxy is about 100,000 light-years or 30.7 Kpc and its radius is about 15 Kpc.

- Studies in 1999 found that the dark matter halo of a spiral galaxy extends about 10 to 20 times the size of the visible regions (slide #3). Using a factor of 15, the radius of the dark matter halo would extend to perhaps 225 Kpc.

- Thus, some 10^{16} eV protons with a Larmor Radius of 11 Kpc might stay within the Milky Way galaxy's 15 Kpc radius. A 10^{17} eV proton with a Larmor Radius of 110 Kpc might remain within the halo's 225 Kpc radius, but a 10^{18} eV proton would probably leave both the galaxy and the halo.

Page 93 of *How Dark Matter Created Dark Energy And The Sun*

SLIDE #86

Only One Dark Matter Candidate Establishes the Approximate Size of the Milky Way

- From slides #85 and #86 it has been shown that the Larmor Radius for a 10^{16} eV proton in the Milky Way's extragalactic magnetic field of 10^{-9} gauss is 11 Kpc; for a 10^{17} eV proton it is 110 Kpc; and for a 10^{18} eV proton it is 1,100 Kpc.

- In slide #17 entitled, "Cosmic-Ray Energy Distribution for the Milky Way," the "knee" of the curve falls on a proton energy of approximately 5×10^{15} eV, which means that some protons with that energy might stay within the galaxy.

- From the above it could be concluded that the Milky Way galaxy should have a minimum radius of about 11 Kpc, compared to astronomers' estimate of about 15 Kpc.

- This seems to provide additional evidence that UHE protons are a credible dark matter candidate. No other currently proposed dark matter candidate can be used to estimate the size or even the order of magnitude of the size of the Milky Way galaxy.

Page 94 of *How Dark Matter Created Dark Energy And The Sun*

SLIDE #87

Drexler's Cosmic-Ray Cosmology
Applied to Galaxy Formation

Some Plausible Speculations

- When a UHE proton moving along a certain path crosses orthogonal magnetic field lines, it is deflected up or down depending upon the magnetic field direction. The degree of deflection is proportional to the orthogonal magnetic field strength.

- Any magnetic deflection of a UHE proton reduces its velocity in the direction of its original path for two reasons: the deflection itself will cause its direction to change, and synchrotron radiation losses will reduce its forward velocity.

- Thus, if UHE protons that are moving through space at a certain velocity encounter a bulge (increase) in the magnetic field strength, they will not pass through the magnetic bulge region as quickly as through a no-bulge region and could possibly linger in the region. To describe this magnetic attraction effect for relativistic protons and electrons, I will use the term "attract" in quotes.

Page 95 of *How Dark Matter Created Dark Energy And The Sun*

SLIDE 88

Drexler's Cosmic-Ray Cosmology
Applied to Galaxy Formation

Some Plausible Speculations

- Since electrons lose energy through synchrotron radiation at a rate 11 trillion times faster than protons, electrons would lose their energy quickly and tend to circulate and accumulate in magnetic-bulge regions.

- Such electron-filled regions would add a coulomb attractive force to UHE protons, slowing them down and also facilitating their conversion into hydrogen atoms. Conceivably, this process could lead, in some cases, to negatively charged proto-galaxies or galaxies surrounded by positively-charged UHE proton dark matter. This, in turn, could lead to lightning-like proton electrical discharges creating a multitude of gamma-ray bursts (GRBs) of immense proportions.

- As the UHE protons slow down in close proximity to a magnetic-field bulge they could linger close by, adding mass and gravitational attraction to the magnetic bulge region.

Page 96 of *How Dark Matter Created Dark Energy And The Sun*

SLIDE #89

Drexler's Cosmic-Ray Cosmology
Applied to Galaxy Formation

Some Plausible Speculations

- Magnetic-field bulges "attract" relativistic electrons and protons. The relativistic proton and electron "attraction" to a magnetic bulge may be different from, but as real as, the gravitational attraction between two masses.

- Previously, my concept of the "death spiral" was discussed with regard to UHE protons plummeting into the galaxy from the halo. In that case the galaxy magnetic field was about 2,000 times the strength of the extragalactic field. The Milky Way is an example of a significant magnetic-field bulge covering a large area in space that should capture almost all UHE protons spiraling through. In the solar system, the Sun is an example of a very significant magnetic-field bulge. With its magnetic field strength of about 50 gauss and the Milky Way's magnetic field strength at 2×10^{-6} gauss, the magnetic field ratio is 25 million.

Page 97 of *How Dark Matter Created Dark Energy And The Sun*

SLIDE #90

APPENDIX C

Presented here are Chapters 44-48, 50, and 56, the Epilogue, of J. Drexler's May 22, 2006 book *Comprehending and Decoding the Cosmos*. It is the second book of Drexler's astro-cosmology trilogy. These chapters are referred to as Chapters 2006-44, 2006-45, etc.

CHAPTER 2006-44

Cosmic DM Mystery #15
Astrophysical Emergence Of Dark Matter Halos, After Eons

Astrophysical Emergence Of Dark Matter Halos And Long, Large, Dark Matter Filaments Could Place Constraints On The Identity Of Dark Matter Particles

The author would like to propose *astrophysical emergence* of dark matter halos as the 15th Cosmic DM Mystery of the Universe, which leads to the following rhetorical question: What type of dark matter particles could facilitate, expedite, or explain the formation of dark matter halos around galaxies and galaxy clusters through *astrophysical emergence?*

Is it possible that a simple microscopic entity or particle in enormous quantities could create, over an evolution period of hundreds of millions to a few billion years, dark matter halos around galaxies and galaxy clusters as well as long, large, dark matter filaments through the principle of *astrophysical emergence* or *emergent evolution?* Astrophysical emergence or emergent evolution or emergent behavior is involved when an enormous number of simple microscopic entities or particles interact with each other strongly so as to form a large entity of complex behavior through collective self-organization. The complex behavior of the resulting large entity would be unpredictable and unprecedented and would represent cosmic bodies or systems or Cosmic DM Mysteries with a high level of sophistication and evolution. These various characteristics are typical of *emergence.*

It is widely believed that *emergence* or emergent evolution of small, simple entities into a large entity of complex behavior requires that the simple pre-existing entities or particles be highly interactive with each other. Galaxy-orbiting relativistic protons meet this requirement.

Furthermore, the complex behavior of any such large entity would not be a known property of the simple microscopic entity or particle, nor could the complex behavior be deduced or predicted from the properties of the simple entity or particle. By relating *astrophysical emergence* to the formation of dark matter halos around galaxies and galaxy clusters and also to the formation of long, large, dark matter filaments that form galaxy clusters where they intersect, constraints can be placed on the nature of the simple microscopic entity (a dark matter particle) to help identify it.

When one deals with the concept of *emergence*, normally the focus is on the collective complex behavior of the resultant large, sophisticated entity because the characteristics of the simple entity or entities, present in enormous quantities, are known already. In those cases, the interest is in the unpredictable and unprecedented complex behavior of the large, sophisticated system that evolves.

The author has a different interest than that of traditional *emergence*. Most readers already are familiar with many of the various forms of complex behavior of the cosmic bodies and Cosmic DM Mysteries of the Universe such as galaxies, dark matter halos, galaxy clusters, the long, large filaments of dark matter, etc. In this case, the author is trying to identify the simple entity (or entities) that could have led to the collective self-organization and the complex behavior exhibited by these various cosmic bodies, systems, or Cosmic DM Mysteries.

If there is a principal simple entity or particle that collectively evolves into emergent behavior, it must have some special interactive characteristics. In the case of a hurricane, there are many simple entities interacting, such as pressure, temperature, moisture, wind, etc.

However, in *astrophysical emergence* involving dark matter halos, there are only a limited number of possible simple entities or particles that could qualify because, for example, they must also represent 80% to 85% of the mass of the Universe.

For example, even if the CDM WIMPs do exist, it is difficult to imagine that large numbers of *weakly interacting* massive particles (only through gravitation) could interact strongly enough with each other to exhibit *emergent* behavior — namely, collective self-organization into a large cosmic body (such as the long, large filaments of dark matter) with complex behavior that is unpredictable and unprecedented.

On the other hand, the enormous numbers of relativistic protons interacting collectively, over an evolution period of hundreds of millions to billion years, possibly could form the DM halos and the long, large, DM filaments that form galaxy clusters where they intersect. These DM filaments could represent one steady-state *emergence* solution. Following the crashing intersection of two DM filaments, new boundary conditions could be established that possibly could lead to the formation of spherical DM halos as a second steady-state *emergence* solution.

The relativistic protons individually follow a number of well-known principles of physics, which probably could lead to collective self-organization and complex behavior after an interaction period of hundreds of millions to a few billion years because of their strong and various coulomb, electromagnetic, and nuclear interactions listed below:

1. Relativistic protons create magnetic fields.

2. Relativistic protons are deflected (accelerated) by magnetic fields.

3. Relativistic protons follow almost-circular spiral paths that are determined by the Larmor Radius equation in direct proportion to the kinetic energies of the relativistic protons and inversely proportional to the strength of the local orthogonal magnetic field.

4. Relativistic protons emit synchrotron radiation and, as a result, lose kinetic energy and slow down, which leads to the emission of even more synchrotron radiation.

5. Relativistic protons moving from one level of orthogonal magnetic field to a higher one will experience an increase in synchrotron radiation.

6. When the velocity of a relativistic proton is reduced through synchrotron radiation losses or collisions with dust or atoms, relativistic mass is also reduced, thereby lowering its gravitational strength.

7. Collisions of relativistic protons with dust, molecules, or atoms will create pions that quickly decay into muons that, in turn, decay into electrons and thereby form a high-velocity plasma comprising protons, helium nuclei, electrons, and muons.

These seven characteristics of relativistic protons should play an important role in *astrophysical emergence*. If one is trying to explain the creation of DM halos and the long, large, DM filaments through the concept of *astrophysical emergence* utilizing either relativistic protons or weakly interacting theoretical WIMPs as possible simple microscopic entities, the relativistic protons with their strong and multiple interactive nature would be a far more promising candidate.

CHAPTER 2006-45

Cosmic DM Mystery #16
UHECRs Arrive At Earth From Galaxy Superclusters

Ultra-High Energy Cosmic Ray Protons Arriving At Earth Probably Departed From A Galaxy Supercluster Or A Massive Galaxy Cluster

See SigChar A, C, D, G, M, and Chapters 46 and 47.

The author would like to propose ultra-high energy cosmic ray protons departing from massive galaxy clusters toward the Earth as the 16th Cosmic DM Mystery. The follow-up rhetorical question is: What type of dark matter particle could lead to ultra-high energy cosmic ray protons departing for Earth from massive galaxy clusters?

On April 22, 2005, astrophysics paper astro-ph/0504512[37] was posted by Drexler on the Physics arXiv. This astrophysics paper predicted that "The DM [dark matter] halos in and around the Virgo Supercluster probably would contain protons with energies above 6×10^{18} eV and probably would be the departing location of the highest energy cosmic rays arriving at the Earth." On July 29, 2005, astrophysics paper astro-ph/0507679 entitled, "Massive galaxy clusters and the origin of Ultra High Energy Cosmic Rays"[38] was posted on the Physics arXiv by astronomers Elena Pierpaoli (California Institute of Technology) and Glennys Farrar (New York University). Their paper reported:

> We find an excess of highest energy cosmic rays (above 50 EeV or 5×10^{19} eV) which correlate with massive galaxy cluster positions within an angle of about one degree. The observed correlation has a chance probability of order 0.1%.

Drexler's paper is entitled "Identifying Dark Matter Through the Constraints Imposed by Fourteen Astronomically Based 'Cosmic Constituents'"[37]. Although this 19-page paper discusses 14 Cosmic Constituents, (now referred to as "Cosmic DM Mysteries") of the Universe that may be created or influenced by or have a special relationship with possible dark matter candidates, only one of them involves the departure location of Earthbound ultra-high energy cosmic rays (UHECRs). This particular Cosmic DM Mystery is best described by the following rhetorical question: What type of dark matter particles could create the first "knee" at 3×10^{15} eV, the second "knee" between 10^{17} eV and 10^{18} eV, and the "ankle" at 3×10^{18} eV of the cosmic ray energy distribution near the Earth? See Appendix B, Slide #17.

After determining that galaxy-orbiting *relativistic proton dark matter* best satisfies the constraints imposed by the 14 astronomically based Cosmic Constituents/Cosmic DM Mysteries, Drexler was then able to use this information to (1) explain the probable reasons for the two "knees" and the "ankle" of the cosmic ray energy distribution at the Earth and (2) identify four probable departing locations of Earthbound cosmic ray protons representing four different proton energy levels and associated proton flux levels.

This led to Drexler's conclusion in his April 22, 2005 paper that the departing location for Earthbound cosmic ray protons with energies greater than 6×10^{18} eV probably would be the dark matter halo in and around the Virgo Supercluster. In considering Pierpaoli and Farrar's July 29, 2005 paper, "Massive galaxy clusters and the origin of Ultra High Energy Cosmic Rays"[38], note that a "massive galaxy cluster" would be an accurate description of the Virgo Supercluster; that about 90% of cosmic ray nuclei are high-velocity protons; and that *ultra-high energy (UHE)* cosmic ray protons, also known as UHECRs, are defined as being at or above the 10^{18} eV energy level.

From the theoretical prediction of this book's author in April 2005 and from the actual astronomical detection several months later, it would appear that ultra-high energy cosmic ray protons arriving at Earth probably had departed from one or more massive galaxy clusters.

The explanations presented in this chapter for Cosmic DM Mystery #16 and the sources of the UHECRs are augmented by explanations and information presented in Chapter 47, for which Chapter 46 is a prerequisite.

CHAPTER 2006-46

Cosmic DM Mystery #17
Starburst Galaxies Form Via Merging Galaxy Clusters

The Merging Of Spiral Galaxy Clusters Create Starburst Galaxies That Exhibit Star Formation Rates (SFRs) As Much As 50 Times Higher Than The SFRs Of Spiral Galaxies

See SigChar G, J, K, S, T, W, and X.

The author designates the above-titled phenomenon as the 17th Cosmic DM Mystery of the Universe, which then leads to the rhetorical question: What type of dark matter particle could cause, expedite, facilitate, or explain the starburst galaxy phenomenon?

Starburst galaxies show evidence of a transient increase in star formation rate by a factor as much as 50. Most starburst galaxies are associated with merging spiral galaxy clusters. The starburst phenomenon may be galaxy-wide or limited to a region of the galaxy such as the galaxy nucleus.

An excellent source of basic astronomical information about starburst galaxies is a six-page article, "Starburst Galaxies"[39] by astronomer William C. Keel of the University of Alabama, posted on the University of Alabama's website: http://www.astr.ua.edu/keel/galaxies/starburst.html.

Starburst galaxies, with SFRs rising by as much as a factor of 50, are created when two clusters of spiral galaxies merge. Starbursts are associated with spiral galaxies that have been disturbed from their steady-state condition. The starbursts are usually confined to a few hundred parsecs from the nucleus of a spiral galaxy, although some starbursts occur throughout the galaxy disk. Large quantities of dust are also associated with starburst galaxies as well as the blue stellar emission from the young stars. Much of the star formation appears to be associated with very compact star clusters of about one hundred million stars in a region of a few parsecs in diameter near the galaxy nucleus.

A majority of spiral galaxies that are found in close pairs, known as *interacting* galaxies, demonstrate an increase in SFRs from about 30% to is as much as 100%.

The four preceding paragraphs describe the principal features and characteristics of starburst spiral galaxies and interacting spiral galaxies. The next step is to utilize this information in conjunction with the relativistic proton dark matter theory/cosmology to extract a plausible explanation for the starburst phenomenon.

In the dark matter halo around a normal spiral galaxy, the lowest energy relativistic protons are close to and penetrate the surface of the enclosed galaxy, while the more than an order of magnitude higher energy relativistic protons would be orbiting at the outer diameter of the dark matter halo at radii more than an order of magnitude greater than those in close proximity to the galaxy. Therefore, there is normally no interaction between a galaxy's hydrogen atoms and molecules and the highest energy relativistic protons in the outer diameter of its dark matter halo.

However, if two spiral galaxies along with their relativistic dark matter halos are interacting or merging, the highest energy relativistic protons in the dark matter halo of one galaxy could bombard the hydrogen atoms and hydrogen molecules of the other galaxy. Further, if two spiral galaxy *clusters* are merging, the ultra-high energy protons orbiting the galaxy *clusters*, if disturbed from their steady-state orbits, can smash into the hydrogen atoms and molecules of individual spiral galaxies. The enormous number of muons generated by these UHECR collisions would be capable of catalyzing the hydrogen fusion reactions associated with the well-known starburst galaxy phenomenon.

If these last two paragraphs are not fully understood, it is suggested that the SigChar references at the beginning of this chapter be reviewed. SigChar T describes a low star formation rate dwarf galaxy where there is a physical gap between the galaxy disk and the "hollow" core of the halo, leading to a low star formation rate. It also explains how a normal star formation rate is associated with the disk of a spiral galaxy overlapping its dark matter halo "hollow" core, thereby permitting more and higher energy relativistic protons to enter the galaxy and generate muons where the hydrogen atoms and molecules are located. Applying these same concepts to closely situated interacting pairs of spiral galaxies, it is not surprising that their star formation rates rise about 30% to 100% above those of normal spiral galaxies.

Why should starburst galaxies have a factor of 50 higher star formation rate compared to the 30% to 100% rise for close and interacting pairs of spiral galaxies? There may be two reasons. Starburst galaxies are usually formed from the merging of two spiral galaxy *clusters*. The ultra-high energies of DM relativistic protons in the DM halo of a galaxy *cluster* are probably about 30 times higher than those in a DM halo of a single spiral galaxy. (See Chapter 50.)

These ultra-high energy DM protons orbiting galaxy *clusters* have the potential to produce very high SFRs. Also, there should be a large reservoir of hydrogen atoms and molecules to ignite into stars near the spiral galaxy nucleus because gravitational tidal forces will move such slow-moving atoms and molecules toward the gravitationally attractive black hole over millions to billions of years.

The hydrogen atoms and molecules near the nucleus of a spiral galaxy would not normally participate in new star formation. Photographs of spiral galaxies typically exhibit the blue color of a new star formation at the outer periphery of those galaxies in the spiral arms. Therefore, it would not be surprising that in a starburst galaxy involving merging galaxy *clusters*, the ultra-high energy protons from the galaxy cluster's dark matter halo could be perturbed from their normal circular orbits because of magnetic field distortions and, thus, could smash into a spiral galaxy straight through to its hydrogen-rich nucleus to ignite new stars. Star ignition could occur through the creation of muons near the galaxy nucleus, followed by particle collisions involving hybrid muonic molecular ions comprised of protons and helium nuclei. Also facilitating this starburst process is the ionization of some of the atomic hydrogen gas in the galaxy nucleus by the ultra-high energy dark matter protons, which speeds up the formation of hydrogen *molecules* and thereby raises the SFR, as explained in Chapters 26 and 41.

This chapter provides a plausible explanation for the starburst galaxy phenomenon, designated Cosmic DM Mystery #17, that cannot be explained by the mainstream theory of star formation where clouds of hydrogen molecules collapse anywhere in a galaxy under their own weight and are heated through compression to hydrogen fusion temperatures.

CHAPTER 2006-47

Cosmic DM Mystery #18
UHECR Protons Via Starburst Galaxies/Merging Galaxies

Spiral Galaxy Clusters, Merging To Form Starburst Galaxies, Were Recently Identified As A Source Of Ultra-High Energy Cosmic Ray Protons

See SigChar A, C, D, G, M, Appendix A, and also Chapters 45 and 46 and reference *39,* which should be considered an integral part of this chapter.

The author designates the above-titled phenomenon as the 18th Cosmic DM Mystery of the Universe, which raises the question: What type of dark matter particle could cause, expedite, facilitate, or explain the creation of ultra-high energy cosmic ray protons by merging galaxy clusters forming starburst galaxies?

In 2005, two research groups independently reported ultra-high energy cosmic rays (UHECRs) emanating from starburst-like galaxies, according to one group, or from merging galaxy clusters, according to the other group. Since starburst galaxies usually involve merging galaxy clusters, it would appear that both research groups could be on similar tracks to the same discovery. On May 13, 2005, Diego F. Torres and Luis A. Anchordoqui posted a paper on the Physics arXiv, astro-ph/0505283 entitled "On The Observational Status Of Ultrahigh Energy Cosmic Rays And Their Possible Origin In Starburst-Like Galaxies". [40]

Then, on July 29, 2005, Elena Pierpaoli and Glennys Farrar posted a paper on the Physics arXiv, astro-ph/0507679 entitled "Massive galaxy clusters and the origin of Ultra High Energy Cosmic Rays" [38] in which the "massive galaxy clusters" are described as "a merging pair of clusters". In their paper, Pierpaoli and Farrar suggest a possible explanation for the observed phenomenon as follows:

> A merging pair of clusters would be expected to have very large scale, strong magnetic shocks which could be responsible for accelerating UHECR even if there is no AGN [active galactic nucleus] or GRB [gamma ray burst] associated with the galaxy clusters.

Note that Pierpaoli and Farrar believe that lower-energy cosmic ray protons are accelerated into UHECRs through magnetic shocks created in the merging galaxy *clusters*. The paper by Torres and Anchordoqui similarly concludes that lower energy cosmic ray protons are probably accelerated to ultra-high energy UHECRs within the merging starburst-like galaxies.

Perhaps both research groups are correct in concluding that the UHECRs may have been accelerated. However, there is another possibility. During the pre-merger period, UHECRs, defined as having energies at or above 10^{18} eV, might have been orbiting galaxy *clusters* within their dark matter halos in a steady-state manner according to the Larmor Radius equation. Given the general size of galaxy clusters and the generally accepted magnitude range of the extragalactic magnetic field, one would conclude that most of the pre-merger orbiting protons in the dark matter halos around the galaxy clusters would be UHECRs.

The galaxy cluster merging process would upset the steady-state Larmor orbiting symmetry of the UHECRs. The combining of the magnetic fields of the two merging spiral galaxy clusters could create transient magnetic field distortions, which would cause a number of UHECRs to be deflected off into space, with some being Earthbound. This theory might be called the deflection-from-orbit theory of

UHECR emission. It is presented as a plausible alternative theory to the shock acceleration UHECR theory, which remains unproven according to the two research groups.

The Pierpaoli and Farrar paper states that the authors have data of UHECRs with energies at 5×10^{19} eV departing from a merging pair of galaxy clusters observed in the SSDS DR3. In the Torres and Anchordoqui paper, the authors indicate the energies of the UHECRs emanating from a starburst galaxy are at or above 10^{18} eV.

Both of these sets of astronomical data appear to be consistent with Drexler's proposed deflection-from-orbit theory of UHECR emission. Therefore, this chapter provides a plausible explanation for the merging-galaxy-cluster-origin-of-UHECRs phenomenon defined by Cosmic DM Mystery #18, as per the Pierpaoli and Farrar paper and possibly as per the Torres and Anchordoqui paper, provided that their starburst galaxy evolved from merging spiral galaxy clusters.

CHAPTER 2006-48

Cosmic DM Mystery #19
Blue Stars In Spiral Arms Vs. Red Stars In Galaxy Nucleus

The Spiral Arms Of Spiral Galaxies Contain Many Hot Blue And Blue-White Stars Less Than One Million Years Old, And In The Galaxy Nucleus There Are Red Stars About Five Billion Years Old

See SigChar A, C, D, G, J, T, W, X, and Chapters 31 and 38.

The author designates the above-stated phenomenon as the 19th Cosmic DM Mystery of the Universe, which raises the rhetorical question: What type of dark matter particle could cause, expedite, facilitate, or explain the creation of the one-million year-old blue and blue-white stars in the spiral arms at the outer diameter of a spiral galaxy, which also has five-billion-year-old red stars located near the galaxy nucleus?

A study of face-on photographs of spiral galaxies clearly shows the many blue and blue-white young stars in the spiral arms at the outer periphery of spiral galaxies. A photograph of the Andromeda galaxy, M31, can be found on the cover of this book. Photographs directed toward the nuclei of spiral galaxies at red wavelengths show old red stars that are estimated to be five billion years old (for Galaxy M81).

These star colors and their locations are widely known. Although there may be generally accepted explanations for the location of the newborn blue and blue-white stars in the spiral arms at the outer periphery of spiral galaxies, the author has not been able to find any published explanation for this preferred star-birth location. This specific star-birth location is not explained by the mainstream theory of star formation where clouds of hydrogen collapse, anywhere in a galaxy, under their own weight and are heated, through compression, to hydrogen fusion temperatures. The mainstream star formation theory provides no clues why spiral galaxies should form their new stars in the spiral arms.

Why are the blue and blue-white stars being formed at the outer diameter of the spiral galaxy disks M81and M31 rather than elsewhere? Could the spherical dark matter halos surrounding spiral galaxies such as M81 and M31, the Andromeda galaxy, play a role in igniting new stars in them?

Let us briefly review the steps in achieving new star formation in an isolated spiral galaxy according to the relativistic proton dark matter star formation theory. This review also explains why the stars forming in an isolated spiral galaxy, surrounded by a relativistic proton dark matter halo, would be located near the outer diameter of the galaxy disk, where the spiral arms are located:

1. An isolated spiral galaxy is surrounded by a spherical dark matter halo comprised of galaxy-orbiting relativistic protons following almost-circular spiral

paths determined by the proton kinetic energies, the local orthogonal magnetic field, and the Larmor Radius equation.

2. Since a spiral galaxy is normally producing stars, its disk must be overlapping the "hollow" core diameter of the dark matter halo. (Spherical dark matter halos surrounding spiral galaxies have "hollow" cores, as explained at the end of Chapter 38.) Relativistic protons near the "hollow" core would lose kinetic energy over time from synchrotron radiation losses and eventually would plunge into the enclosed spiral galaxy and bombard the atomic and molecular hydrogen gas closest to the galaxy surface.

3. The bombarding of the hydrogen gas primarily near the surface by the relativistic protons will produce a number of significant effects. By ionizing some (50% would be optimum) of the atomic hydrogen, the conversion of atomic hydrogen to molecular hydrogen would be accelerated. The protons bombarding the resulting molecular hydrogen would create large numbers of muons, which would catalyze hydrogen fusion reactions. The bombarding protons (and accompanying high-speed helium nuclei) also would trigger nuclear fusion reactions and the ignition of new stars by colliding with the muonic ions created earlier by reactions of muons with the hydrogen and helium.

This scenario provides a plausible explanation for the birth of blue and blue-white stars in the spiral arms near the surfaces of spiral galaxies, as defined by Cosmic DM Mystery #19.

The dark matter relativistic protons also may be performing another role. On one hand, the muons they create can catalyze the hydrogen fusion reaction, thereby igniting new blue and blue-white stars. In addition, the protons add hydrogen to the enclosed galaxy, which can facilitate future star formation and to cause the galaxy to grow by accretion.

The idea of galaxy growth through accretion of baryons provided by the galaxy's dark matter halo is a relatively new idea and, therefore, requires some support from astronomical data. One example that provides such support is the Andromeda galaxy, M31, whose disk seems to have enlarged by accretion by a factor of three over billions of years. See astro-ph/0504164 entitled, "On the accretion origin of a vast extended stellar disk around the Andromeda galaxy."[41]

There is a related point that should be considered. Five billion years ago, the red stars near Andromeda's nucleus might have been blue stars at the outer diameter of a then-much-smaller galaxy. Did the galaxy disk grow in diameter by accretion since then?

A study of 2,800 stars outside Andromeda's disk by the authors of astro-ph/0504164, led to the discovery that these stars were not in the halo of the disk, but actually in an extension of the galaxy disk. From a study of the velocities and directions of the stars in the extended disk, researchers have ruled out the possibility of an earlier merger with another galaxy to explain the tripling of the disk diameter. This would leave hydrogen accretion as the primary source of growth, by a factor of three, in the diameter of the Andromeda disk. The author believes that Andromeda's dark matter halo could have provided the necessary hydrogen or protons for galaxy disk growth as posited by the relativistic proton dark matter theory/cosmology; however, astro-ph/0504164 does not suggest that.

Additional astronomical support for new star formation taking place primarily at the outer periphery of spiral galaxies is found in the empirical Schmidt law that posits that the SFR of isolated spiral galaxies is highly correlated with their average *molecular hydrogen surface density.*

See Chapter 53, which links the Schmidt law to the relativistic proton dark matter theory/cosmology and thereby provides (1) a plausible explanation for the Schmidt law, (2) support for the relativistic proton dark matter theory and cosmology, and (3) support for Cosmic DM Mystery #19, the formation of new stars in the spiral arms near the surfaces of spiral galaxies.

The star formation phenomena defined by Cosmic DM Mystery #19 cannot be explained by the generally accepted mainstream theory of star formation where clouds of hydrogen molecules collapse anywhere in a galaxy under their own weight and are heated through compression to hydrogen fusion temperatures.

CHAPTER 2006-50

Cosmic DM Mystery #21
Different Dark Matter For Small Galaxies And For Clusters

**Two Different Types Of Dark Matter
Halo Particles Reported for Smaller Galaxies
And For Galaxy Clusters.
On August 26, 2005, Researchers Reported
That In Low-Surface Brightness Galaxies And Dwarf
Galaxies, The Self-Interaction Cross Section Of Dark
Matter Particles Appears To Be High; While For More
Massive Systems Such As Galaxy Clusters,
The Self-Interaction Cross Section Of Dark Matter
Particles Appears To Be Low.
No Explanatory Theory Was Offered.**

See SigChar C, D, G, M, S, and Chapter 44.

The author selects the above-stated dual-dark-matter phenomenon to be the 21st Cosmic DM Mystery of the Universe, which raises the question: What type of dark matter particle could have a high self-interaction cross section for DM halos of low surface brightness galaxies and dwarf galaxies, but a lower self-interaction cross section for DM halos of spiral galaxy clusters?

On August 26, 2005, Bo Qin, Ue-Li Pen, and Joseph Silk posted a four-page paper, "Observational Evidence for Extra Dimensions from Dark Matter"[42] on the Physics ArXiv as astro-ph/0508572. They pointed out that the cold dark matter theory was questioned in the late 1990s because numerical simulations indicated a cuspy core, which was never observed astronomically. They then gave credit to David N. Spergel and Paul J. Steinhardt of Princeton University, the authors of astro-ph/9909386, "Observational evidence for self-interacting cold dark matter"[43], for proposing in 1999 a self-interacting cold dark matter mode that could possibly solve the cuspy core problem.

However, astro-ph/0508572 points out that in the year 2000, numerical simulations:

> ... found that the [self-interaction] cross sections needed to produce good agreement with galaxies turned out to produce galaxy clusters too large or too round to be consistent with observation.

Bo Qin, Ue-Li Pen, and Joseph Silk then came to a very significant conclusion:

> These lines of evidence might suggest that, instead of being a fixed value, the dark matter self-interaction cross section as proposed by Spergel and Steinhardt, is likely to vary in different dark matter systems — cross sections whereas more massive systems (like galaxy clusters) tend to have smaller cross sections.

These researchers then provided an equation proposed by others that estimates the self-interaction cross section of dark matter particles. They then added the following:

> The nature of this self-interaction between dark matter particles is unknown. Its strength generally must be put in by hand. We note that the scattering cross section in Eq.(1) decreases with increasing velocity, which is characteristic of long-range forces (like gravity or Coulomb forces).

The above-quoted three paragraphs essentially describe the characteristics of the relativistic protons in the relativistic proton dark matter theory. Chapter 49 points out that the proton energies in the Milky Way's dark matter halo probably range in energy between 1×10^{16} eV and 2×10^{17} eV, while Chapter 16 points out the proton energies in the dark matter halo surrounding the Local Group cluster of galaxies probably range in energy levels between 3×10^{17} eV and 6×10^{18} eV, or about 30 times higher.

Note that this book is dealing with the coulomb forces of the protons and, therefore, the self-interaction or interaction cross section declines with proton velocity or as the square root of energy. Using this information and the two previous paragraphs about spiral galaxies and their clusters, one can arrive at a roughly estimated factor of about five higher self-interaction cross section for the relativistic protons in the DM halos of dwarf and LSB galaxies, as compared with the relativistic protons in the DM halos of spiral galaxy clusters.

Thus, the relativistic proton dark matter theory appears to be compatible with the dual-dark-matter phenomenon defined by Cosmic DM Mystery #21, for which a plausible explanation has been provided in this chapter. This is an indication that *dark matter relationism* seems to be applicable to Cosmic DM Mystery #21.

Chapter 44 indicates the need for the dark matter particles to be highly self-interactive in order that *emergent evolution* principles could lead to the formation of dark matter filaments and dark matter halos around galaxies/clusters during an evolution period of millions to a few billion years.

CHAPTER 2006-56

EPILOGUE

The Local Group's Dwarf Spheroidal Satellite Galaxies Help Define Dark Matter

After a draft manuscript of this Book II of the trilogy had been completed, research astronomers at Cambridge University reported at a press conference on February 3, 2006 that their research on 10 dwarf spheroidal (dSph) satellite companions of the Milky Way and Andromeda revealed surprising information about dark matter.

Researcher Professor Gerry Gilmore, according to the press, reported that:

1. Dark matter particles are not slow and cold, but instead appear to be moving at 9 kilometers per second and have an apparent temperature of about 10,000° C, which is higher than the 6,000° C at the surface of the sun.

2. "The strange thing about dark matter is that it doesn't give off radiation."[51]

3. "There must be some form of repulsion (between the dark matter particles)..."[52] "We have to start looking into the physics of the interactions between dark matter particles — not just at the way they respond to gravity."[50]

4. "This indicates that dark matter clumps together in building blocks which have a minimum size," said team member Dr. Mark Wilkinson in an article by Zeeya Merali in NewScientist.com on February 6, 2006, entitled, "'Tepid' temperature of dark matter revealed."[50] "This is 1,000 light years across, with 30 million times the mass of the sun," said Dr. Wilkinson.

In BBC NEWS, 6 February 2006, in an article entitled, "Dark matter comes out of the cold," Jonathan Amos reported that researchers have been able to establish that the galaxies [studied by Cambridge researchers] contain about 400 times the amount of dark matter as they do ordinary matter.

Dr. Wilkinson, a member of Professor Gilmore's team, was further quoted in the February 6, 2006 NewScientist.com article as saying, "No matter what size, how bright, or how many stars they contained — all the galaxies seemed to be sitting in roughly the same amount of dark matter."[50] Does Wilkinson's statement imply that the dark matter halos existed prior to the formation of the enclosed galaxies?

In the same article, Dr. Robert Minchin, an astronomer at the Arecibo Observatory in Puerto Rico, is quoted as saying, "It [the results] contradicts the WMAP findings, which suggested that warm dark matter was unlikely."[50]

In the Education.Guardian, an article was published by science correspondent Alok Jha on February 6, 2006, entitled "Research into dwarf galaxies starts to unlock the deep secrets of dark matter".[51] A bullet title reads, "1,000-light-year-wide bricks make up the universe".

In a February 10, 2006 article in the journal Science, entitled, "Dwarf Galaxies May Help Define Dark Matter,"[52] Daniel Clery reported that Gilmore suggests that they [the dark matter particles] interact with one another to spread out evenly. "There must .be some form of repulsion [between the dark matter particles]..."

In National Geographic News, an article was published by James Owen on February 13, 2006, entitled, "Dark Matter properties 'Measured' for the First Time, Study Says," which quotes Professor Gilmore. Each galaxy studied was found to contain the same amount of dark matter. "That was a big surprise," Gilmore added. "The galaxies contain very different numbers of stars, so like everybody else we thought they would cover a very wide range of masses, but they don't. There seems to be a minimum mass for a galaxy." Do these statements by Gilmore imply that the dark matter halos existed prior to the formation of the enclosed galaxies?

On the morning of February 23, 2006, Professor Gilmore gave an invited progress report on his dark matter research by video conferencing at "Dark Matter 2006", the 7th UCLA Symposium on Sources and Detection of Dark Matter and Dark Energy in the Universe. The principal change from his previous data was that his recent estimate for the minimum dark matter mass is 50 million solar masses compared to 30 million solar masses estimated previously.

Since no research paper has been published on the Cambridge University findings, these phenomena have not been designated a Cosmic DM Mystery even though the Cambridge University three-year research program's results appear to be significant. The further confirmation of the following five different types of astronomical data could provide strong support for the Drexler dark matter theory/cosmology and could weaken support for the cold dark matter theory:

1. The measurements of the high-velocity/high-temperature dark matter particles.

2. The lack of thermal radiation from the 10,000° C dark matter particles.

3. Indications of strong self-interactions between dark matter particles.

4. The minimum size and mass of the dark matter clumps.

5. The contradiction of the earlier WMAP findings, which have provided support for the cold dark matter model.

Professor Gilmore's statement, "The strange thing about dark matter is that it has temperature but it does not give off radiation," raises a question that may be answered by the relativistic proton dark matter theory. The relativistic cosmic ray protons that enter the Earth's atmosphere are very fast and have very high energy but have low entropy and, therefore, would exhibit very little thermal radiation. Of course, when they decelerate, they radiate photons as when they enter a magnetic field or collide with atoms, molecules, photons, or dust, but only a portion of this radiation would be thermal radiation. Thus, Professor Gilmore's above statement essentially describes low-entropy relativistic dark matter protons and, therefore, his statement seems to provide support for the relativistic proton dark matter theory.

Professor Gilmore's statement, "We have to start looking into the physics of the interactions between dark matter particles — not just at the way they respond to gravity," has been addressed in different ways in Chapter 44 and Chapter 50 of this book, and the mutual repulsion of positively charged protons is long known.

Chapter 44 is entitled "Astrophysical Emergence Of Dark Matter Halos And Long, Large, Dark Matter Filaments Could Place Constraints On The Identity Of Dark Matter Particles". As pointed out in that chapter, the author believes that without the strong self-interaction of the dark matter particles, the dark matter halos around spiral galaxies and their galaxy clusters would not have formed under *astronomical emergence* or *emergent evolution* principles.

Chapter 50 is entitled "Two Different Types Of Dark Matter Halo Particles Reported For Smaller Galaxies And For Galaxy Clusters". Chapter 50 relates that researchers have discovered that the self-interaction of dark matter particles orbiting galaxy clusters is much weaker than the self-interaction of dark matter particles orbiting LSB and dwarf galaxies. The relativistic proton dark matter theory is utilized to explain the difference.

Dr. Mark Wilkinson's (and Professor Gilmore's) statements that all dark matter halos around galaxies have roughly the same amount of dark matter, support Drexler's relativistic proton dark matter theory. This theory posits (see Jerome Drexler's April 2005 paper, astro-ph/0504512,[37] pages 12-

13) that, "Dark matter halos around galaxies and galaxy clusters have outer diameters and 'hollow' core diameters determined by the galactic and extragalactic magnetic fields and the energy spectrum of the relativistic protons." The Drexler paper further states, "Also, the author believes that the outer diameter and core size diameter of DM halos are not significantly affected by the amount of galaxy mass enclosed."

The discovery by the Cambridge researchers that dark matter is found in minimum-size building blocks of 1,000 light-years across is consistent with the earlier discovery of the diminutive galaxy I Zwicky 18 (see Chapter 39), which is just 3,000 light-years across.

One possible explanation for the minimum size dark matter building blocks of 1,000 light-years across will now be presented in the following five paragraphs, based on Drexler's relativistic proton dark matter theory:

Synchrotron radiation (or emission) is electromagnetic radiation that is emitted by charged particles moving at relativistic velocities in circular/spiral orbits in an orthogonal magnetic field. The rate of synchrotron emission is inversely proportional to the product of the radius of curvature of the particle's orbit and the fourth power of the mass of the particles. The radius of curvature of the particle orbits is determined by the Larmor Radius equation. The Larmor

Radius is directly proportional to the kinetic energy of the particles and inversely proportional to the magnetic field orthogonal to the paths of the particles.

The 10 dwarf spheroidal satellite galaxies of the Milky Way/Andromeda studied by Gilmore and his group all lie relatively close by their host galaxy. For example, the Fornax spheroidal galaxy is approximately 138 kpc from the Milky Way. (Dwarf irregular galaxies lie much further away from their host galaxies.)

Thus, the two magnetic fields that influence the radius of curvature of the proton paths in the dark matter halos of a dwarf spheroidal satellite galaxy are the magnetic field strength of its host galaxy (2×10^{-6} gauss for the Milky Way) and the much lower extragalactic (intergalactic) magnetic field strength (estimated at 1×10^{-9} gauss).

Therefore, in the vicinity of the DM halo of the dwarf spheroidal galaxy (dSph) 138 kpc from the Milky Way, the combined magnetic field strength probably would be above 1×10^{-8} and less than 1×10^{-7} gauss. This magnetic field strength may indicate that the DM halo of the dSph would be more than one order of magnitude smaller in diameter than that of the Milky Way's DM halo. Also, the average synchrotron radiation losses of the dSph's DM halo's protons may be more than one order of magnitude greater than the average synchrotron radiation losses for protons in the DM halo of the Milky Way.

These synchrotron radiation losses of the halo protons of the dSph galaxies, in turn, would lower the average kinetic energy of the orbiting dark matter protons, which would further lower the orbiting protons' radius of curvature. This, in turn, would further increase the synchrotron radiation energy losses of the orbiting protons. These energy losses eventually could accelerate and destroy the stability of the protons' orbital paths, thereby providing some evidence that relativistic proton dark matter halos of the dSph galaxies may not be able to exist stably below some minimum size radius and related minimum mass.

The newly discovered characteristics of the dark matter particles described by the Cambridge University researchers and mentioned in this chapter appear to have plausible explanations based upon the Relativistic-Proton dark matter.

REFERENCES

References for *Discovering Postmodern Cosmology*, the third book of the trilogy.

1. F. Zwicky, 1937 *Astrophys. J.* (Lett) 86, 217

2. V. C. Rubin, N. Thonnard and W. K. Ford, 1978 *Astrophys. J.* (Lett) 225 , L107

3. V. Rubin, *Bright Galaxies – Dark Matters* (Amer. Inst. Physics, New York, 1997), p. 109-116

4. G. Blumenthal, S. Faber, J. R. Primack, and M. J. Rees, 1984 *Nature* **311**, 517

5. J. Drexler, *How Dark Matter Created Dark Energy and the Sun* (Universal Publishers, Parkland, Florida, USA, 2003)

6. J. Drexler, *op. cit.*, p. 18

7. A. S. Bishop, *Project Sherwood – The U.S. Program in Controlled Fusion* (Addison- Wesley Publishing Company, Inc., Reading, Massachusetts, U.S.A. 1958) p. 177-178

8. J. Drexler, *Comprehending and Decoding the Cosmos* (Universal Publishers, Boca Raton, Florida, USA, 2006)

9. P. Berardelli, "Dark Matter Not a Done Deal?", Science NOW Daily News 31 October 2007

10. S. Hawking, Lecture – Life in the Universe,
http://www.hawking.org.uk/lectures/life.html

11. S. Koppes, The University of Chicago Chronicle, Vol. 24, No.5, November 18, 2004,
http://chronicle.uchicago.edu/041118/entropy.shtml

12. J. Drexler, 2005, astro-ph/0504512 v1

13. J. Drexler, 2007, physics/0702132

14. http://en.rian.ru/analysis/20060620/49784754.html
Russian UV/EUV Space Telescope 2010

15. C. H. Gibson, 2006, astro-ph/0606073

16. Sir Martin Rees, "Cosmology" on TV Channel 14:
http://www.channel4.com/science/microsites/W/what_we_still_dont_know/findoutmore.html

17. R. Lieu and J. Mittaz, *Soft X-Ray Emission From Clusters of Galaxies and Related Phenomena* (Kluwer Academic Publishers, Norwell, MA, USA2004)

18. NASA: "Motions in nearby galaxy cluster reveal presence of hidden superstructure," NASA Marshall Space Flight Center Release No. 04-231, Huntsville, Alabama, September 9, 2004, p.1

19. NASA – Bullet Cluster news release:
http://www.nasa.gov/home/hqnews/2006/aug/HQ_062 97_CHANDRA_Dark_Matter.html

20. NASA – Hubble Repair Mission:
http://www.nasa.gov/home/hqnews/2006/oct/HQ_06343_HST_announcement.html

21. F. A. Aharonian, et al, 2006, astro-ph/0610509 v2, "H.E.S.S. observations of the Galactic Center region and their possible dark matter interpretation"

22. "Gamma-ray glow bathes Milky Way"
http://findarticles.com/p/articles/mi_m1200/is_n19_v152/ai_19997827

23. F. A. Aharonian, et al, 2006, Nature 439, 695-698, "Discovery of very-high-energy Gamma rays from the Galactic Centre ridge"

24. M. J. Jee, et al, 2007, arXiv: 0705.2171, "Discovery of a Ringlike Dark Matter Structure in the Core of the Galaxy Cluster Cl 0024+17"

25. G. R. Farrar and R. A. Rosen, 2006, arXiv: astro-ph/0610298, "A New Force in the Dark Sector?"

26. L. Wai, 2007,arXiv:astro-ph/0701884v1, " Dark Matter Searches with Glast"

27. F. Bournaud, "Missing Mass in Collisional Debris from Galaxies", Science 25 May 2007, Vol.316 no.5828, p.1166-1169

28. S. McGaugh, "Seeing Through Dark Matter", Science 3 August 2007, Vol.317., no.5838, p. 607- 608

29. NASA – "Dark Matter Mystery Deepens in Cosmic 'Train Wreck'" August 16, 2007
http://www.nasa.gov/mission_pages/chandra/news/07-090.html

30. NASA – Dark Matter Introduction
http://imagine.gsfc.nasa.gov/docs/science/know_l1/dark_matter.html

31. L. Gao and T. Theuns, "Lighting the Universe with Filaments", Science 14 September 2007, Vol. 317, no. 5844, p. 1527-1530

32. Michael J. Disney, "Modern Cosmology: Science or Folk Tale?"
http://www.americanscientist.org/template/AssetDetail/assetid/558 39?&print=yes

33. J. R. Brownstein, J. W. Moffat, "The Bullet Cluster 1E0657-558 evidence shows modified gravity in the absence of dark matter" MNRAS Volume 382 Issue 1 p. 29-47, November 2007 http://www.blackwell-synergy.com/doi/abs/10.1111/j.1365-2966.2007.12275.x

34. Pierre Auger Collaboration, "Correlation of the Highest-Energy Cosmic Rays with Nearby Extragalactic Objects", Science 9 November 2007, Vol.318, no. 5852, p. 938-943

35. Cosmic Inflation – From Wikipedia:
http://en.wikipedia.org/wiki/Cosmic_inflation

36. P. Ricon, BBC, "Hubble makes 3D dark matter map"
http://news.bbc.co.uk/2/hi/science/nature/6235751.stm

37. M. S. Turner, 2001, arXiv::astro-ph/0108103, "Dark Energy and the New Cosmology"

38. Is Dark Matter a Source of High Energy Gamma Rays? http://www.physorg.com/news84797343.html

39. A. Mahdavi, et al, 2007, arXiv:0706.3048 v1, A Dark Core in Abell 520

40. D. F. Torres and L. A. Anchordoqui, 2005, astro-ph/0505283 v1 "On the Observational Status of Ultrahigh Energy Cosmic Rays and their Possible Origin in Starburst-Like Galaxies."

41. E. Pierpaoli and G. Farrar, 2005, astro-ph/0 507679 v3, "Massive galaxy clusters and the origin of Ultra High Energy Cosmic Rays"

42. C. A. Scharf, D. R. Zurek, and M. Bureau, 2004, astro-ph/0406216, The Chandra Fornax Survey - I: The Cluster Environment

43. R. Panek, 2007, "Out There"
http://www.nytimes.com/2007/03/11/magazine/11dark.t.html?_r=1&oref=slogin

44. A. Cho, "Untangling the Celestial Strings", Science, 4 January 2008, Vol. 319. no.5859, p. 47-49

45. F. Nicastro, S. Mathur, and M. Elvis, "Missing Baryons and the Warm-Hot Intergalactic Medium", Science, 4 January 2008, Vol. 319. no.5859 p.55-57

46. "NASA Announces Details of Hubble Servicing Mission" Jan.8, 2008,
http://www.reuters.com/article/pressRelease/idUS169689+08-Jan-2008+PRN20080108

47. R. A. Ibata and G. F. Lewis, "The Cosmic Web in Our Own Backyard", Science, 4 January, 2008, Vol.319. no. 5859 p. 50-52

BIBLIOGRAPHY
AND SUGGESTED SOURCES

The following is a list of books and articles that the author suggests as recommended reading. Readers should not assume that any author listed below or throughout this book agrees with the author's views or theories.

J. N. Bahcall and J. P. Ostriker, *Unsolved Problems in Astrophysics* (Princeton Univ. Press, Princeton, New Jersey, 1997).

A. S. Bishop, *Project Sherwood – The U.S. Program in Controlled Fusion* (Addison-Wesley Publishing Company, Inc., Reading, Massachusetts, U.S.A. 1958).

R. Clay and B. Dawson, *Cosmic Bullets – High Energy Particles in Astrophysics* (Helix Books, Addison-Wesley, Australia, 1998).

K. Croswell, *The Alchemy of the Heavens* (Anchor Books, Doubleday, New York, 1995).

J. Drexler, *How Dark Matter Created Dark Energy And The Sun* (Universal Publishers, Parkland, Florida, USA, 2003).

J. Drexler, *Comprehending and Decoding the Cosmos* (Universal Publishers, Boca Raton, Florida, USA, 2006).

D. Filkin, *Stephen Hawking's Universe* (Basic Books, New York, 1997).

M. W. Friedlander, *Cosmic Rays* (Harvard Univ. Press, Cambridge, MA and London, England, 1989).

M. W. Friedlander, *A Thin Cosmic Rain – Particles From Outer Space* (Harvard Univ. Press, Cambridge, MA and London, England, 2000).

H. Friedman, *The Astronomer's Universe* (W.W. Norton & Company, New York, 1998).

T.K. Gaisser, *Cosmic Rays and Particle Physics* (Cambridge University Press, Cambridge U.K., 1990).

D. Goldsmith, *The Astronomers* (St. Martin's Press, New York, 1991).

A. H. Guth, *The Inflationary Universe* (Helix Books, Perseus Books, Reading, Massachusetts, 1998).

R. P. Kirshner, *The Extravagant Universe – Exploding Stars, Dark Energy and the Accelerating Cosmos* (Princeton Univ. Press, Princeton, New Jersey, 2002).

E. W. Kolb and M. S. Turner, *The Early Universe* (Addison-Wesley, USA, 1990).

L. Krauss, *Quintessence* (Basic Books, A Member of the Perseus Books Group, New York, N.Y., 2000).

T. S. Kuhn, *The Structure of Scientific Revolutions* (The University of Chicago Press, Chicago Illinois, 1970).

R. B. Laughlin, A Different Universe (Basic Books, A Member of the Perseus Books Group, New York, N.Y., 2005).

M. S. Longair, *High Energy Astrophysics,* Volume I, Second Edition (Cambridge Univ. Press, Cambridge, UK 1992).

M. S. Longair, *High Energy Astrophysics*, Volume II, Second Edition (Cambridge Univ. Press, Cambridge, UK, 1994).

M. S. Madsen, *The Dynamic Cosmos* (Chapman & Hall, London, England, 1995).

V. Rubin, *Bright Galaxies – Dark Matters* (Amer. Inst. Physics, New York, 1997).

G. Rudiger and R. Hollerbach, *The Magnetic Universe – Geophysical and Astrophysical Dynamo Theory* (Wiley-VCH Verlag GmbH & Co. KGaA, Weinheim, Germany, 2004).

D. W. Sciama, *Modern Cosmology and the Dark Matter Problem* (Cambridge Univ. Press, Cambridge, UK, 1993).

J. Silk, *A Short History of the Universe* (Scientific American Library, a Division of HPHLP, New York, 1997).

P. Sokolsky, *Introduction to Ultrahigh Energy Cosmic Ray Physics* (Westview Press, A member of the Perseus Books Group, Boulder Colorado, USA, 2004).

T. Stanev, *High Energy Cosmic Rays* (Springer-Verlag Berlin Heidelberg New York, 2004)

T. X. Trinh, *The Secret Melody* (Oxford Univ. Press, New York, 1995).

J. N. Wilford, *Cosmic Dispatches – The New York Times Reports on Astronomy and Cosmology* (W. W. Norton & Company, New York and London, 2002).

W. S. C. Williams, *Nuclear and Particle Physics* (Oxford Univ. Press, Oxford and New York, 1991).

S. Yoshida, *Ultra-High Energy Particle Astrophysics* (Nova Science Publishers, Inc., New York N.Y., 2003).

GLOSSARY

Accelerating Cosmos: Accelerating expansion of universe.

Accretion: An infall of matter on an object.

AGASA: Akeno Giant Air Shower Array.

AGN: Galaxies having active galactic nuclei.

Alpha Particle: The nucleus of a helium atom.

Andromeda Galaxy: Twin galaxy of the Milky Way. The two galaxies comprise most of the Local Group's mass.

Astronomical Unit (A.U.): The average distance from the Sun to the Earth, equal to 149,598,000 kilometers.

Astrophysics: The study of the composition and other physical properties of celestial objects.

Astrophysical Emergence: See emergence and emergent evolution.

Astrophysical Dynamo Effect: Relativistic protons orbiting galaxies will create magnetic fields through the astrophysical dynamo effect under which the relativistic protons moving in Larmor orbits create magnetic fields. These same magnetic fields in turn determine the proton paths, eventually reaching a steady-state solution for the magnetic fields and the proton paths after an emergent evolution period involving millions to billions of years.

"Attract"/"Attraction": A new term (in quotes) that refers to the movement in space of a UHE relativistic proton from one magnetic field strength to a higher magnetic field strength, resulting in the slowing down and dwelling in the region of the higher magnetic field.

Baryon/Baryonic: An elementary particle that is subject to the strong nuclear interaction. The proton and neutron and combinations of them are baryons.

Big Bang: The cosmological theory that holds that all the matter and energy in the Universe was concentrated in an immensely hot and dense point, which exploded 13.7 billion years ago. Beginning 2008, see Chapter 12.

Black Hole: An object that exerts such enormous gravitational force that nothing, not even light or other forms of electromagnetic radiation, can escape from it.

BL Lac, BL Lacertae Objects: Objects that are in the category of AGN's, now known as blazars, that exhibit no emission lines but have a continuum emission from radio frequencies through X-ray frequencies.

Bottom-Up Theory: The theory that small galaxies form first and larger galaxies are formed through mergers of small galaxies.

Bremsstrahlung Radiation: Braking radiation of a proton or other charged particle.

CERN: Center for European Nuclear Research.

Chandra X-ray Observatory: Part of NASA's fleet of "Great Observatories" along with the Hubble Space

Telescope. Chandra allows scientists to obtain unprecedented X-ray images of exotic environments.

Closed Universe: A universe in which the density of matter is greater than the critical density and that should thus collapse onto itself in the future.

Cluster Soft Excess (CSE): EUV and soft X-ray emission from a galaxy cluster beyond what would be expected based upon its temperature.

COBE: Cosmic Background Explorer.

Cold Dark Matter (CDM): Non-baryonic matter consisting of elementary particles of relatively high mass that are moving relatively slowly. (The term "cold" indicates a low temperature and thus a small energy of motion.)

Collective Self-organization: A term very similar in meaning to emergence and emergent evolution.

Collision Cross Section: A measure of the probability that an encounter between particles will result in the occurrence of a particular atomic or nuclear reaction.

Coma Cluster: A galaxy cluster that contains about 1,000 galaxies. The gravitational effects of dark matter were discovered in this galaxy cluster.

Comet: A body of ice and dust, with a nucleus of typically about 10 kilometers in diameter.

Copernicus' (Nicolaus) Concept: The Earth rotates daily on its axis and the planets revolve in orbits around the Sun. He authored *On The Revolutions of Heavenly Spheres* published about 1542.

Cosmic DM Mysteries/Cosmic Constituents: Dark matter mysteries or unexplained phenomena regarding celestial bodies or cosmic matter such as their shape, mass distribution, particle abundance ratios, dimensions, density, location, maturity, acceleration, velocity, linear momentum, angular momentum, particle energies, star rotation curves, hydrogen fusion reactions, particle energy distributions, particle transformations, star ignition, and star formation rates.

Cosmic Inflation: The hyper rapid expansion of the Universe a fraction of a second after the Big Bang. See Chapter 31.

Cosmic Microwave Background (CMB) Or Cosmic Background Radiation (CBR): The microwave radiation that bathes the entire Universe and that dates from the epoch when the Universe was just 300,000 years old.

Cosmic Origins Spectrograph (COS): An ultraviolet (UV and EUV) spectrograph scheduled to be installed on the Hubble telescope in August 2008.

Cosmic-ray Cosmology: A new term to describe a cosmology recently developed by J. Drexler, based upon UHE protons and cosmic ray protons.

Cosmic Rays: Particles (mostly protons and electrons) that have been accelerated somewhere in the Universe to very high energies.

Cosmic Web: The cosmic web is considered to be the framework on which the universe is built. It is comprised primarily of dark matter that makes up about 83 percent of the mass of the universe. It is explained in Chapter 33.

Cosmology: The study of the Universe as a whole, and of its structure and evolution.

Cosmos: An orderly, harmonious, and systematic Universe.

Coulomb Force: The force between two coulomb charges or electrically charged particles.

CSE: See Cluster Soft Excess.

Dark Energy (as defined in the past): A hypothetical form of energy that permeates all space and has negative pressure resulting in a repulsive gravitational force. The accelerating expansion of the Universe has been attributed to dark energy. Beginning 2008, see Chapter 21.

Dark Galaxy: A galaxy with a total light output or luminous level from stars below an established minimum threshold level. Such galaxies represent extreme cases of low surface brightness (LSB) galaxies.

Dark Matter (as defined in the past): Matter that is detected only by its gravitational pull on visible matter. The composition has been unknown; it might consist of very low mass stars or supermassive black holes, but Big Bang nucleosynthesis calculations limit the amount of such baryonic matter to a small fraction of the critical mass density. If the mass density is critical, as predicted by the simplest versions of inflation, then the bulk of the dark matter must be a gas of weakly interacting non-baryonic particles, sometimes called WIMPs (Weakly Interacting Massive Particles).

Deuterium: A chemical element whose nucleus consists of a proton and a neutron, created mainly in the first three minutes of the Universe's history.

DOE: U.S. Department of Energy.

Doppler Effect: The variation in the energy and color of light caused by the motion of a source of light relative to an observer. If the source is receding, the energy decreases and the light is shifted toward the red. If the source is approaching, the energy increases and the light is shifted toward the blue.

Doppler Shift: The shift in the received frequency and wavelength of an electromagnetic wave that occurs when either the source or the observer is in motion. Approach causes a shift toward shorter wavelengths and higher frequencies called a blue shift. Recession has the opposite effect, called a red shift. The expansion of the Universe causes ancient electromagnetic wave emissions to exhibit a doppler red shift.

Dwarf Galaxy: A galaxy with a small size and mass. The average diameter is about 15,000 light-years; that is, about one-sixth of that of a normal galaxy. Masses range from 100 million to 1 billion solar masses, about 1,000 to 10,000 times less than the mass of an ordinary galaxy. Dwarf galaxies may be spheroidal or irregular, but dwarf spiral galaxies have not been observed.

Dwarf Spheroidal Galaxy (dSph): A dwarf galaxy that is spheroidal in shape, has an old stellar population, and lies close to a large host galaxy as a satellite.

Dwarf Irregular Galaxy (dIrr): A dwarf galaxy that is irregular in shape, has a young stellar population, and is a satellite of a large host galaxy, but lies further away from its host galaxy than would a dwarf spheroidal galaxy.

EeV: 10^{18} eV.

Electromagnetic Wave: A pattern of electric and magnetic fields that moves through space. Depending on the wavelength, an electromagnetic wave can be a radio wave, a microwave, an infrared wave, a wave of visible light, an ultraviolet wave, a beam of X rays, or a beam of gamma rays.

Electron: The lightest of the subatomic particles with electrical charge. The electron has a mass of 9×10^{-28} kilograms and is negatively charged.

Electron Volt (eV): The energy released when a single electron passes through a one-volt battery.

Elliptical Galaxy: A galaxy observed as an oval-shaped system generally composed of old stars, a large black hole, and containing little or no gas and dust.

Emergence, Emergent Evolution: A theory that new characteristics and qualities appear in the evolutionary process at more complex organizational levels (than that of the pre-existent entities such as a molecule, a cell, or a particle) and which cannot be predicted solely by studying less complex levels of organization but which are determined by a rearrangement of pre-existent entities.

Entropy: A measure of the unavailability of a system's energy to do work.

ESO: European Space Organization.

EUV: Extreme ultraviolet radiation.

Extragalactic, Intergalactic: The regions of the Universe outside of any galaxy.

Field Galaxies: The variety of galaxy types typically found in galaxy surveys.

Galactic Disk: A flattened aggregation of stars, gas, and dust in a spiral galaxy. The average disk is some 90,000 light-years in diameter and 300 light-years thick. In the Milky Way, the stars complete one turn around the galactic center every 250 million years, at a velocity of 230 kilometers per second.

Galactic Halo: A spherical region around a spiral galaxy populated by old stars and globular clusters. Observations suggest that it is surrounded by a dark matter halo some 10 to 20 times larger than the galaxy and more massive.

Galaxy: A system of stars (10 million in a dwarf galaxy, 100-200 billion in an average galaxy like the Milky Way, 10 trillion in a giant galaxy) held together by gravity.

Galaxy Cluster: A dense grouping of several thousand galaxies bound by gravity, with an average diameter of some 60 million light-years, and an average mass of a few million billion solar masses.

Gamma Ray: An electromagnetic wave with a wavelength in the range of 10^{-13} to 10^{-10} meters, corresponding to photons with energy in the range of 10^4 to 10^7 electron volts. Their energies are higher than X rays.

Gauss: A measure of the strength of a magnetic field.

General Relativity: A gravitational theory proposed by Albert Einstein in 1915, which is more accurate than that of Newton. The two theories differ mainly in situations where gravitational fields are very intense, such as around a pulsar or black hole. General relativity constitutes the theoretical support of the Big Bang theory.

GeV: G stands for Giga, or 10^9. Thus, GeV is one billion electron volts.

Gravitational Field: A field of force surrounding a body of finite mass. The field of force is defined as the force that would be experienced by a standard mass positioned at each point in the field.

Gravitational Tidal Force: The tidal force responsible for attraction between all matter. The weakest of the four forces, gravitational force possesses the longest range.

GRB: A gamma ray burst.

Group of Galaxies: A collection of about 20 galaxies held together by gravity, some six million light-years across and averaging between one and 10 trillion solar masses.

Gyr: Gigayear, or one billion years.

GZK Cosmic-Ray Cutoff: A theory limiting proton energies. According to the currently questioned 1966 Greisen-Zatsepin-Kuzmin (GZK) cutoff theory, protons with energies greater than 6×10^{19} eV would interact with the cosmic microwave background radiation and lose energy through radiation and thus would not travel more than 50 Mpc, or about 160 million light-years. In 1998, Coleman and Glashow wrote a paper entitled, "Evading the GZK Cosmic-Ray Cutoff," which showed that for very high energy cosmic rays, the GZK cutoff would not apply. See Chapters 29 and 31.

Halo: The region around a galaxy that contains dark matter and some halo stars.

Helium: A chemical element with a nucleus of two protons and two neutrons (helium-4). A second, far-less-abundant isotope has two protons and one neutron (helium-3).

HESS High Energy Spectrographic System: A system used for detecting gamma rays.

Hubble Law: The law discovered in 1929 by the American astronomer Edwin Hubble, which states that the distance of galaxies varies in proportion to their red shift and, thus, because of the Doppler effect, to their velocity of recession. The law gave birth to the idea of an expanding universe.

Hydrogen: The lightest of all chemical elements, consisting of one proton and one electron. Hydrogen makes up 75% of the mass of the Universe.

Isotropy/Isotropic: The property of the Universe to be similar in every direction.

Kpc: The abbreviation for a kilo parsec where a parsec equals 3.26 light-years.

Kepler's laws: Laws concerning the motions of the planets in their orbits derived by Johannes Kepler in the 16th century.

Large Magellanic Cloud (LMC): The larger of two irregularly shaped galaxies closest to the Milky Way located in the far southern sky and visible to the unaided eye.

Larmor Radius (for a proton): A proton crossing an orthogonal/magnetic field and entering into a spiral path. The radius of a cycle of that spiral path is called the proton Larmor Radius for that cycle.

$$\text{Proton Larmor Radius} = 110 \text{ Kpc} \times \frac{10^{-8} \text{ gauss}}{B} \times \frac{E}{10^{18} \text{ eV}}$$

Law of Conservation of Linear Momentum: The total linear momentum of the mass objects in a group remains unchanged. See momentum.

Light-year: The distance traveled by light (which moves at a velocity of 300,000 kilometers per second) in one year and equal to 9,460 billion kilometers.

Local Group: A grouping of galaxies extending over a region of space of about 10 million light-years, of which the Milky Way and Andromeda are the principal and most massive members (one trillion solar masses each). It also includes dwarf galaxies.

Low-Surface Brightness (LSB) Galaxy: A diffuse galaxy with a surface brightness that is one magnitude lower than the ambient night sky.

Magnetic Bulge: A significant rise in the orthogonal magnetic field experienced by a relativistic proton.

Magnetic Field: A field of force in space, created by a magnet or by an electric current, that guides the trajectories of electrically charged particles by exerting an electromagnetic force.

Mass: The measure of the inertia of an object, determined by observing the acceleration when a known force is applied. An object with mass creates a gravitational field, which is defined in this glossary. When a proton travels at relativistic velocities, it has a relativistic mass equal to its energy divided by the square of the speed of light.

Maxwell's Equations: A set of differential equations describing space and time dependence of the electromagnetic field and forming the basis for classical electrodynamics.

Microwave: An electromagnetic wave with a wavelength of between one millimeter and 30 centimeters.

Milky Way: The galaxy to which our solar system belongs, whose central regions appear as a band of light or "milky way" that we can see from Earth in clear night skies.

Missing Baryons: About 4.5 percent of the mass/energy of the universe is estimated to be baryonic mass. It can be confirmed as of 10 billion years ago, but half seems to be missing now. See Chapter 32.

Missing Mass: An outmoded name for the dark matter of the Universe.

MNRAS: Monthly Notes of the Royal Astronomical Society.

m_o: The symbol m_o representing the mass of a proton when it is not moving (the rest mass).

Momentum: The linear momentum of an object, equaling the product of its mass and velocity. If no external forces are acting on a group of mass objects, the Law of Conservation of Linear Momentum requires that the total linear momentum of the mass objects in the group remains unchanged.

Muon (contraction of the earlier mu-meson; taken as a symbol for meson, and used to distinguish it from the short-lived pi-meson): An unstable elementary particle that belongs to the lepton family, that is common in the cosmic radiation near the Earth's surface, that has a mass about 207 times the mass of the electron, and that exists in negative and positive forms.

Muonic Ion: Two nuclei of atoms in close proximity, usually one of them being a proton and the other being a deuteron, a helium nucleus, or another proton, being orbited very closely by a single negative muon weighing 207 times as much as an electron at rest. Muonic ions are best known for catalyzing low temperature nuclear fusion reactions.

NASA: National Aeronautics and Space Administration.

Neutralino: A theoretical non-baryonic particle, which is an amalgam of the superpartners of the photon (which transmits the electromagnetic force), the Z boson (which transmits the so-called weak nuclear force), and perhaps other particle types. Although the neutralino is heavy by normal standards (at least 35 times the mass of a proton), it is generally thought to be the lightest supersymmetric particle.

Neutron: A subatomic particle with no electric charge, one of the two basic constituents of an atomic nucleus.

Nucleosythesis: The production of a chemical element from hydrogen nuclei or protons, as in stellar evolution.

Ockham's Razor Logic (also called Occam's razor): A scientific and philosophic rule that the favored explanation for an unknown phenomenon is the simplest of the competing theories. It should be preferred to the more complex, or that explanations of unknown phenomena be sought first in terms of known quantities rather than through assumptions.

NSF: National Science Foundation.

Open Universe: A Universe in which the density of matter is less than the critical density and which will thus expand forever.

Oort Cloud: A region in the outer limits of the solar system where billions to trillions of comets reside.

Orthogonal: Intersecting or lying at right angles.

Parsec: An astronomical unit of distance equal to 3.26 light-years or approximately 19 trillion miles.

Pion (contraction of pi-meson): A short-lived meson that is primarily responsible for the nuclear force and that exists as a positive or negative particle with mass 273.2 times the electron mass or a neutral particle with mass 264.2 times the electron mass.

Population I Stars: A younger generation of stars with ages from a few million years to about 10 billion years and with a relatively large fractional abundance (about 1% of mass) of elements heavier than helium. The Sun is in this category.

Postmodern Cosmology or Postmodern Big Bang Cosmology: A recently discovered cosmology in which the output of the Big Bang is comprised almost entirely of relativistic protons and helium nuclei in a ratio of about 12:1 and dark matter is comprised of the identical subatomic particles in the essentially the same ratio.

Primordial Ripples: The mass perturbations in the early Universe that may have evolved into galaxies.

Proto-galaxy: A cloud of gas and ions that is evolving into a galaxy.

Proton: A positively charged particle composed of three quarks that, together with the neutron, forms atomic nuclei. The proton is 1,836 times more massive than the electron.

RBB: Relativistic Big Bang. See Chapters 12, 14, and 35.

Recombination: In the traditional theory, between 300,000 and 700,000 years after the Big Bang, the plasma of free electrons and hydrogen nuclei that condensed to form a neutral gas, in a process called recombination. The prefix "re" is not meaningful here, however, since according to the Big Bang theory, the electrons and protons (hydrogen nuclei) were combining for the first time ever.

Red Shift: A shift to longer wavelengths and lower frequencies, typically caused by the Doppler effect in a receding object or caused by the expansion of the Universe.

Reductionism: A procedure or theory that reduces or attempts to reduce complex data or phenomena to simple elements or terms. It is an inward-looking approach that in physics usually means a search for subatomic particles in an attempt to understand or explain some unusual phenomenon.

Relationism: An analytical procedure, method, concept, or theory, developed by Jerome Drexler, that attempts to identify dark matter by determining which cosmic phenomena (called Cosmic DM Mysteries) may be facilitated, expedited, influenced by, or have a special relationship with dark matter. This outward-looking cosmological concept is used to determine the nature and characteristics of dark matter's influence on and relationship with cosmic phenomena as a means of DM identification.

Relativistic-Baryon Dark Matter: Formerly known as Relativistic-Proton Dark Matter or relativistic-proton dark matter. The ratio of protons to helium nuclei ranges between 10:1 and 12: 1.

Rotation Curve of a Galaxy: A graph of the orbital velocities of stars or hydrogen as a function of their radial distances from the nucleus of the galaxy radially outward into the surrounding dark matter halo.

R-PDM: Relativistic-Proton dark matter.

Schmidt Law: An empirical law, for isolated spiral galaxies, of the correlation between the star formation rate (SFR) and the overall average molecular hydrogen surface density.

Second Law of Thermodynamics: The entropy (disorder) of the Universe always increases over time.

SFR: Star formation rate.

Signature Characteristics (SigChar): An extensive list, for each dark matter candidate, of all the possible features or characteristics of the candidate that can be attributed to it by utilizing any and all laws and principles of physics.

Solar System: The Sun and the objects in orbit around it, which include nine planets, nearly 60 known satellites of the planets, thousands of smaller objects called asteroids, and billions to trillions of comets.

Spiral Galaxy: A flattened, disk-like system of stars and interstellar gas and dust with a spherical collection of stars, known as the bulge, at its center. Bright, young stars outline spiral arms in the plane of the disk.

(The) Standard Model: Name given to the current theory of fundamental particles and how they interact.

Star: A sphere of gas consisting of 98% hydrogen and helium and 2% heavy elements in equilibrium under the action of two opposing forces -- the compressive gravity and the outward radiation pressure from the nuclear fusion reactions in its core. The Sun has a mass of 2×10^{30} kilograms, and masses of stars range between 0.1 and 100 solar masses.

Starburst Galaxy: A galaxy experiencing a period of intense star forming activity. They are usually associated with the merging or interaction of two galaxies. This activity may last for 10 million years or more. During a starburst, stars can form at tens, even hundreds, of times greater rates than the star formation rate in normal spiral galaxies.

Subatomic Particles: They could be fast protons or helium nuclei or any fast atom with its electrons removed.

Supercluster: The aggregation of tens of thousands of galaxies held together by gravity and gathered into groups and clusters. Superclusters have the shape of flattened pancakes with an average diameter of 90 million light-years and masses of 10,000 trillion (10^{16}) solar masses.

Supernova/Supernovae: An exploding star, visible for weeks or months, even at enormous distances, because of the tremendous amounts of energy that the star produces. Supernovae typically arise when massive stars exhaust all means of producing energy from nuclear fusion. In these stars, the collapse of the star's core results in the explosion of the star's outer layers. Another type of supernova arises when hydrogen-rich matter from a companion star accumulates on the surface of a white dwarf and then undergoes nuclear fusion. This second type, known as a Type 1a supernova, generates light at a well-known standard level and thus can be used to measure the rate of expansion of the Universe.

Synchrotron Emission or Radiation: Electromagnetic radiation that is emitted by charged particles moving at relativistic speeds in circular orbits in a magnetic field. The rate of emission is inversely proportional to the product of the radius of curvature of the orbit and the fourth power of the mass of the particles. For this reason, synchrotron radiation is not a problem in the design of proton synchrotrons, but it is significant in electron synchrotrons. The synchrotron emission from a relativistic proton or electron is directly proportional to the square of its energy.

TeV: T stands for Tera, or 10^{12}. Thus, TeV is one trillion electron volts.

Tokamak: Tokamak Hydrogen Fusion Test Reactor

Top-Down Theory: The theory that long, large, dark matter filaments of the cosmic web form galaxy clusters where the DM filaments intersect/collide and then galaxies form from the remnants of these collisions.

UHE (Ultra-High Energy) Proton: A proton traveling near the speed of light with an energy of at least 10^{18} eV.

UHECR: Ultra-high energy cosmic ray proton traveling near the speed of light with an energy of at least 10^{18} eV.

Ultraviolet (UV): Ultraviolet light.

Virgo Supercluster: A huge, flattened supercluster that contains the Local Group of galaxies. The Local Group, containing the Milky Way, lies at the edge of the supercluster, while the Virgo Cluster of galaxies is at its center.

White Dwarf Star: A small, dense star with a diameter of about 10,000 kilometers (about the size of Earth) created when a star of less than 1.4 solar masses exhausts the nuclear fuel and collapses under its own gravity. This type of star participates in a Type 1a supernova.

WIMP (Weakly Interacting Massive Particle): The name for a non-baryonic theoretical dark matter candidate that is presumed to have a mass much greater than that of a proton. A neutralino is one form of WIMP.

WMAP (Wilkinson Microwave Anisotropy Probe): A NASA Explorer mission measuring the temperature of the cosmic background radiation over the full sky. This map of the remnant heat of the Big Bang provides data about the origin of the Universe.

X Rays: Electromagnetic radiation with greater frequencies and smaller wavelengths than those of ultraviolet radiation and lower frequencies and longer wavelengths than those of gamma ray radiation.

INDEX

UHE proton, 199, 205, 206,
207, 208, 209, 210, 211,
212, 215, 216, 217, 218, 266
UHECR, xi, 59, 76, 82, 126,
155, 156, 159, 160, 161,
162, 164, 172, 281
Ultraviolet, ix, xi, 24, 27, 28,
29, 30, 31, 32, 43, 47, 48,
51, 64, 74, 93, 97, 98, 114,
115, 182, 266, 269, 281
Universal Publishers, ii, 36, 79,
154, 201, 253, 259
University of Alabama at
Huntsville, 91
University of Chicago, 58, 75,
79, 95, 113, 157, 254, 260
University of Leeds, 158
US Green Card, 84
US Laser Visa, 84

UV, xi, 47, 51, 64, 93, 94, 97,
99, 100, 115, 116, 127, 151,
168, 181, 182, 254, 266, 281
violent radial dispersion, xi,
18, 59, 67, 76, 81, 163, 170,
177
Virgo Supercluster, 281
Wai, L., 255
Watson, Alan, 158, 164
Weakly Interacting Massive
Particle, 23, 65, 68, 184,
267, 281
White dwarf, 280, 281
WIMP, 68, 281
WMAP, 246, 248, 281
XMM-Newton, 29
X-rays, 93, 94, 99, 116
Zurek, D. R., 257
Zwicky, Fritz, 68, 79, 253

Discovering Postmodern Cosmology

Discoveries in Dark Matter, Cosmic Web, Big Bang, Inflation, Cosmic Rays, Dark Energy, Accelerating Cosmos

Learn how a world-class inventor-scientist is currently tackling the greatest scientific mysteries of the universe — and succeeding. With his new book, Drexler provides a viable baseline to jump-start debate on a standard model for *postmodern cosmology*. It is the first book to not only address these seven unsolved cosmic mysteries, shown in this book's subtitle, but also offer plausible explanations for each of them. The correlation of these seven cosmic phenomena by Drexler offers a revolutionary advance in cosmological research and potentially broad acceptance and use of the related concepts.

This book was written for open-minded cosmologists, astronomers, astrophysicists, physicists, engineers, students, enthusiasts and those at NASA, NSF, DOE and ESO who want to understand *postmodern cosmology*. The author's five years of cosmology research, and his successes, convinced him that his *postmodern cosmology* model is correctly based upon the relationships and linkages of these seven cosmic phenomena.

About the Author

Jerome Drexler is the author of three astro-cosmology books, a former Research Professor in physics at NJIT, former Chairman and chief scientist of LaserCard Corp. (Nasdaq: LCRD) and former Member of the Technical Staff of Bell Laboratories. He has been awarded 76 U.S. patents, honorary Doctor of Science degrees from NJIT and Upsala College, a degree of Honorary Fellow of the Technion, an Honorary Life Member of the Technion Board of Governors, an Alfred P. Sloan Fellowship at Stanford University, a three-year Bell Labs graduate study fellowship, the 1990 "Inventor of the Year Award" for Silicon Valley, and the first patent on an optical disk "Laser Optical Storage System".

Printed in the United States
138005LV00002B/94/P